Industrial Safety and Environment

By:

PRATIBHA GARG
(B.E., M.B.A.)
Sr. Lecturer (Mech. Deptt.)
SSIET, Derabassi

ANUPAMA PRASHAR
(B.E., M.B.A.)
Lecturer
CIIS, Mohali

S.K. KATARIA & SONS®
Publisher of Engineering Books
4885/109, Prakash Mahal, Dr. Subhash Bhargav Lane,
Opposite Delhi Medical Association, Daryaganj, New Delhi–110002 (INDIA)
Phone: +91-11-43551243; *Mobile:* +91-9871775858, 7678425209
e-mail: katariabook@gmail.com, katariabooks@yahoo.com
Website: www.skkatariaandsons.com
(e-books, subscriptions and bulk buying also available on our website)

Published by:

S.K. Kataria & Sons®

4885/109, Prakash Mahal, Dr. Subhash Bhargav Lane,
Opposite Delhi Medical Association, Daryaganj, New Delhi-110002 (INDIA)
Phone: +91-11-43551243; *Mobile:* +91-9871775858, 7678425209
e-mail: katariabook@gmail.com, katariabooks@yahoo.com
Website: www.skkatariaandsons.com

First Edition: January, 2005; Second Revised Edition: January 2007,
Reprint: 2026

Laser Typesetting By :
PLUS COMPUTERS, Delhi

Printed at :
Arora Printing Press, Delhi

PREFACE

The purpose of this book is to bring together and integrate in a single text the areas of INDUSTRIAL SAFETY AND ENVIRONMENT. This book is organized in two sections. The first one covers the principle aspects of industrial safety like safety planning, industrial hazards, safety standards, etc. The second section includes detailed discussion on industrial environmental factors like lighting, noise, ventilation, heat stress, humidity etc. We have attempted to make each chapter independent and self contained however, for use in any order or combination.

The author wish to acknowledge the fact that the development and publication of this book would have been possible with the encouragement of our colleagues, the challenge and motivation of our students, and the support and understanding of our families.

Pratibha Garg, Anupama Prashar

CONTENTS

CHAPTER 1

INTRODUCTION TO SAFETY

✍ **The question is not whether We will die, but how we will live.**
— Borysenko, Joan

1.1. SAFETY

Safey is the state of being SAFE

S = spot the hazard
A = assess the risk
F = Find a safer way
E = everyday

So, safety is the state of being certain that adverse effect will not be caused by some agents under defined conditions.

Any method or technique or process which can minimize unwanted events (accidents) in an industrial concern may be referred to as a method, technique or process of Industrial safety.

Industrial safety is the science and art of identifying, evaluating and controlling workplace hazards. It includes measure to prevent human exposures to chemical and physical agents as well as faulty or unsafe work practices.

It is connected with the safety of people's life and property as well as the general situation of development and stability. Safety is not a matter of chance, it is something that the employees and their families have a right to expect from their employers. There is no such thing as rank when its comes to safety. *It is not "us" and "them" its "partnership".*

The importance of industrial safety was realized because every year millions of industrial accidents occur which result in either death or in temporary or permanent disablement. Latest figure estimate workplace accident costs the country at least Rs. 280 million each year. This is the economics cost only. As well there is the human cost in terms of emotional sufferings to the partners, children, family and workmates of those killed at work. This is immeasurable.

The road toll (casualties in road accidents) is often well published but workplace death and injury toll is often overlooked. For many organizations health and workplace safety is a compliance cost and they try and get away with what they see as bare *minimum*.

Health and safety are basic desire and instinct.

Industrial safety is mainly concerned with minimizing hazards. Hazard is a state physical or chemical having the potential to injure the person, or impairemement of health.

Risk or danger arises out of hazard. Risk is the product of two factors: that is probability of an event which might occur and severity of the event if it occurs. While analyzing risks, the severity of the events that might occur is more important then the probability of the event to occur.

In other words:

RISK = PROBABILITY OF THE EVENT THAT MIGHT OCCUR × SEVERITY OF THE EVENT

In the above formula the probability of each event that might occur may remain constants, whereas the severity of each event that might occur varies and therefore when analyzing risk we are more concerned about the severity of the event that might occur and not the probability of the event that might take place.

1.2. SAFETY AND PRODUCTIVITY

Productivity is a measure of effectiveness of the use of resources to produce goods and services. Safety and productivity are closely related to each other. To improve productivity in an organization workplace must be conducive.

The following points or situations reveal the relation between safety and productivity:

(*a*) If an organization is free from hazards and accidents then the white collar employees *i.e.*, managerial staff can save their valuable time otherwise lost in the following activities.
- Assisting injured workers.
- Arranging for the production of an injured worker to be continued by some other worker.

- Preparing accident report.
- Selecting and training a new worker to replace the injured person.
- Investigating the cause of accident.

Therefore, the managerial staff will possibly be able to perform their managerial functions viz planning, organizing, staffing, directing, controlling, co-coordinating etc. more smoothly which in turn will possibly boost the productivity of the organization. On the other hand if serious hazards and accidents occur in an organization at an alarming rate then the managerial staff will not possibly be able to perform their managerial functions smoothly and as a result productivity of the organization will decrease.

(*b*) If the workers are mentally ensured that they are working in a safe working environment free from hazards and their life and limbs are safe in all possible aspects then they will possibly work with more zeal and concentration and thus the quality and productivity will increase considerably. On the other hand if workers feel that their working environment is adverse to safety then their morale will become low and will show little interest in their work and thus the productivity will decrease considerably.

(*c*) If in an industrial organization the workplace is safe then in such situations the workers will be highly motivated to perform their jobs in an efficient manner and thus the productivity will increase considerably. On the other hand if workplace is not safe then the workers will possibly be demotivated to put their best efforts and thus the productivity will decrease considerably.

(*d*) If the working environment is safe then the labour turnover and absenteeism will be reduced and each and every worker will possibly be optimistic and will possibly contribute their best efforts which in turn will help improve productivity. On the other hand if working environment is unsafe then the workers will not contribute their best efforts which will increase the absenteeism and thus the productivity will decrease considerably.

(*e*) If the working place is safe then there is considerable reduction in accidents and hence reduction in direct and indirect cost of accidents. Damage to machines, materials, buildings is also reduced in safe work environment. Thus there is an improvement in productivity of the organization. On the other if work environment is unsafe then the occurrence of accidents and consequent damage to machines, materials, buildings will be increased and hence the productivity will get decreased.

(*f*) If the working conditions in an organization are conducive to safety then accidents are a far cry in such situations and each and every worker will possibly be optimistic will possibly contribute their best efforts and expertise to achieve optimum and or expected production and thus the productivity will increase on the other hand.

If working conditions in an organization are not conducive to safety then it might possibly lead to accidents. Accident of a particular worker lowers the morale of the fellow workers. The fellow workers in such situation becomes pessimistic and obviously become aware of hazardous nature of their job and consequently fellow workers will not contribute their best efforts to achieve expected production and thus productivity will decrease.

1.3. EMPLOYEES PARTICIPATION IN SAFETY

One of the keys to successfully promote safety in an organization is to involve employees. No safety strategy can be successful unless the employees willingly participate in it. The employees usually know better than anyone else where hazard exist. Since safety is a matter of direct interest to the employees, naturally it can be expected that they would like to participate.

> *A fundamental rule of management is 'if you want employees to make a commitment, involve them from the start'.*

One of the most effective strategies for getting employees to commit to the safety programme is to involve them. Emphasis on the participation of the employees should be more in the safety programme than in any area of management because employees are the ones who must follow safety rules. Various measure can be adopted to create the interest of employees to participate in safety, some of them have been found to be very effective. These are:

- **Safety Committees**
- **Safety Campaigns**
- **Safety habits**

Safety Committees

In most modern industries, safety committees have become almost a permanent feature of life. These committees usually consist of members from the management, the supervisor and the workers. Normally the members are rotated. In large industrial organizations, there may be many departmental sub-committees working under the main committee.

The representatives of such committees working in town, state and national safety councils, whose main tasks is to discover potential hazards, and to find out whether there are careless workers, whether workers use safety devices and precautions. They also collect safety data and call for safety suggestions from workers and educate them in the various aspects of safety in a department of plant.

Safety Campaigns

These are essential in certain industries in which the accident, hazards are more serious. These make the workers 'safety conscious'. Safety campaigns are the responsibility of safety committees and safety officers of the company. In such campaigns, the emphasis is on developing an adequate warning system, particularly when there is possibility of a hazard. Posters and slogans can be used very effectively in many situations. Some safety related slogans:

> **'Use safety goggles in welding areas'**
> **Use head covers while going through this passage'**
> **'Safety consciousness will pay you rich dividends'**
> **'No smoking or igniting matches in this area'**

In transport industries, posters or signs can be used that substantially leads to road safety. Posters or signs that warn in advance, tell the precautions one should take, remind one to be cool and careful and reinforce positive safety behaviours are effective. The posters which are designed to create a panic among the workers are not as effective as simple, directly understandable commands.

A scared person is less likely to be safe in his behaviour. Sometimes safety use exhibitions, documentaries and other propaganda material are used to enlighten workers on matters of safety. Such occasional props for stimulating employee's interest in safety consciousness are of great value.

Safety Habits

Most industrial operations are safe if they are performed in a right way without clumsiness or awkwardness. There are correct and efficient methods of doing things which must be so well mastered that there is never a chance for bodily movements which are unsafe. The inexperienced workers are most likely to be involved in accidents. They make unsafe movements because of their work habits. If employees are not taught the right ways of performing their operations, they may acquire some unsafe habits which would expose them to accidents. It is in this sense that right type of training acquires a great importance in accident prevention. If good working habits are ingrained in the workers at the time of their training, they become an inseparable part of their behaviour, and they observe the necessary safety precautions without repeated warnings. It is ncessary, therefore, to place enough emphasis on safety training when entrants in the organization are trained in their respective trades.

1.4. SAFETY AND PLANT LAYOUT

Plant layout is a technique by which location of machines, the plant services as well as the processes within the factory premises are so arranged that the greatest possible high quality output coupled with lowest possible cost of production can be achieved. Proper plant layout aims at minimizing the total operating costs at the same time providing for proper allocation of spaces coupled with systematic arrangement of equipment.

Plant should be designed in manner such that it may provide safety and security to the workers at all costs. A good plant layout provides better working conditions to the workers. It should be designed in a manner such that there is least movement of workers as well as of raw materials coupled with minimum discomfort to all concerned.

An improper plant design can trigger off the accidents. Reduction in the number of accidents should be the aim while designing plants.

A good Plant layout call for high operating efficiency coupled with proper safety provisions, safe and optimum planning of industrial facilities including operating equipments and material handling equipments, personnel etc. Factors like material, workers, circulation or movement space, material handling and storage should be taken care of in plant layout.

While developing design of plant certain engineering techniques like travel and flow process charts, operation charts should be employed in a judicious and systematic manner. Computer aided techniques; mathematical techniques can also help in plant layout.

1.5. SAFETY AND EQUIPMENT DESIGN

The following measures should be taken while equipment design:

- The working equipments should be provided with guards or safety devices.
- In all operations the workers should have an easy and quick access to the controlling mechanism (starting and emergency switches etc.)

- The design and installation of the machine guards and safety devices must be foolproof so far as structural stability, ease of installations and workings and proper specifications etc. are concerned.
- The working equipment (chisels, sledge hammers, hacksaws etc.) which are sharp, pointed ones must be adequately covered to prevent injury.
- The equipment example ladder should be sufficiently strong.

1.6. SAFETY AND WORKING ENVIRONMENT

The working environment includes the atmosphere, temperature and comfort, lighting, air quality, noise and the general surroundings. When any one of these working conditions is substandard they can affects health and safety.

➷ Lighting

There must be levels of lighting sufficient for the work areas and tasks being performed. Too much and too little light can lead to:

- Eye strain
- Headaches
- Blurred vision

IES has given standard of illumination required for different types of work. For example, lighting for routine office work is at a lower lumination level than lighting for jeweler's inspection.

➷ Temperature and Humidity

Very warm or cold work environments can affect the comfort, concentration and productivity of employees. Factors which contribute to comfort include:

- Humidity
- Air movement
- Level of activity
- Clothing

Temperature controls can range from fans and blowers to air conditioning and heating systems. Optimal temperature is between 21 to 24 degrees Celsius. Environments that have extreme temperatures can harm health and should be assessed so that adequate controls can be determined.

➷ Noise

All sounds that are distracting, annoying or harmful to every dry activites such as work, rest, entertainment, study and so forth are regarded as noises. Infact, any sound judged by the recipient as undesirable can be considered as a noise. The noise level has an important bearing on human well-being, even in those cases where it is not perceived consciously. Whether or not a sound is undesired by a person will depend on number of factors, *e.g.*,

(*a*) Loudness

(*b*) Frequency

(*c*) Continuity

(*d*) Variation with time

(*e*) Time of occurrence

(*f*) Information content

(*g*) Origin of sound

(*h*) Recipients state of mind and temperament

(*i*) Background noise level

Noise is generally harmful and a serious health hazard. It has far reaching consequences and has many physical, physiological as well as psychological effects on human being.

Physical Effects: The physcial manifestation of noise pollution is the effect on hearing ability. Repeated exposure to noise may result in temporary or permanent shifting of the hearing threshold of a person depending upon the level and duration of the exposure. The immediate and acute level of noise pollution is improvement of hearing.

Physiological Effects: Headche, increase in rate of heart beat, Narrowing of arteries, muscular strain, nervous breakdown, lowering of concentration are physiological effects of noise.

Psychological Effect: The psychological effects are:

(*i*) Depression and fatigue

(*ii*) Insommia

(*iii*) Straining of senses

(*iv*) Emotional disturbances

1.7. SAFETY MEASURES IN MANUFACTURING ORGANISATION

Manufacturing establishment include all smelters, oil refineries, cement works, mills of every kind, machine and repair shops, addition any other kind or character of manufacturing establishment, of any nature or description whatsoever, whenein any natural product or other articles or materials of any kind, in a raw or unfinished or incomplete state or condition, are converted into a new or improved or different form. The measures taken in manufacturing organization are:

Safety Committee

Safety committee should be constituted in every plant. These committees should consist of members from the management, the supervisor and the workers. All safety programmes should be implemented through safety committee.

Safety Habits

Good working habits should be ingrained in the workers at the time of training, so that these habits become inseparable part of their behaviour, and they observe the necessary safety precautions without repeated warning. It is necessary to place enough emphasis on safety training when entrants in organization are trained in their respective trades.

Maintenance of Plant

Plant should be maintained in manner such that it may provide safety and security to the workers at all costs. All objects which obstruct the passages made for movement of workers should be removed. Passages should not be used to store equipments, goods and materials store rooms with suitable shelves, racks should be provided for keeping electrical and other hazardous equipment.

Protective Clothing

The workers should be provided with protective clothing such as hand gloves, masks, helmets, safety footwear etc. The clothing serve a dual purpose of providing convenience as well as protection at work.

Safety Campaign

Safety campaign make the workers safety conscious. Safety programme should be given a wide publicity through poster and slogans. 'Safety consciousness will pay you rich dividends' and 'No smoking' are example of safety related slogans. Safety contest should be held between plants and the plant with lowest accident rate must be given some award.

Regular Inspection

There should be regular inspection of machines, equipment and electric cables to check any leakage.

Guarding of Machines

Safety guards should be designed and installed to provide positive protection, prevent access to danger zones, avoid inconvenience in operation and give protection against unforeseen contingencies. These are in the form of hard insulated covers provided to the moving or rotating parts of machines.

↳ Material Handling Equipments

Material handling equipments should be installed to carry bulky material form one place to another.

↳ Repair and Maintenance

All doors, stairs, passages shall be kept and maintained in a state of good repair and shall be free of obstruction. No substance which is likely to cause a person to slip, shall be kept near them.

↳ Ergonomics

It deals with physiological and anatomical aspects of man and his interactions with working environment. It is being increasingly used to improve man-machine system. For example equipment, machinery and work procedures should be redesigned to minimise accidents.

1.8. SAFETY AND ECONOMICS

In context of an industry, Safety and Economics are closely related. If the workers are ensured mentally that they working in an organization which is free from hazards and accidents, their life and limbs are safe in all possible manner, their safety at all cost is ensured and accidents are a far cry, then in such situations each and every worker will possibly be optimistic and will contribute their best efforts and expertise to achieve optimum or expected production. Workers will be motivated to perform their jobs in an efficient and skilled manner in a safe working environment. Good working conditions boost the morale of the employees.

In a safe working environment the managerial staff will get more time to perform their managerial functions, *i.e.* planning, organization, staffing, directing, coordinating, controlling etc. smoothly. All these factors boost the productivity of the organization resulting in a strong economics base for the industry.

On the other hand if the working conditions are adverse to safety, then it might possibly lead to accidents at an alarming rate. Workers will feel that they are working in unsafe and hazardous conditions. In such situations the morale of the workers will possibly become low, they will possibly be demotivated to put their best efforts and will show little interest in their assigned job.

The accident of a particular worker lowers the morale of the fellow workers. The fellow workers become pessimistic and become aware of the hazardous nature of the job. They do not contribute their best efforts and expertise to achieve optimum production. The management also do not get enough time to perform their managerial functions smoothly resulting in fall in overall productivity. This inturns weakens the economic base of the organization and the country as a whole.

1.9. ROLE OF GOVERNMENT

The Government of India has established the Factory Advice Service and Labour Institute Bombay, which functions as an integral part to advice on safety matters and enforces safety laws. The Government has drawn up a 'National programme for coordinate Action plan' for control of hazards and protection of health and safety of workers in dangerous manufacturing processes. The action plan lists out the responsibilities of government, management and workers organisations in the field of safety and health in work environment. It also includes model scheme for setting up full safety control system cell in hazardous industries and 'Safety and Health' Accidents Reduction Plan (SAHARA) in all industries.

1.10. NATIONAL SAFETY COUNCIL

The National Safety Council was set up in 1966 to promote safety consciousness, to prevent accidents and to conduct safety programmes lectures. The National Safety Day is celebrated on 4th March every year to mark the foundation day of the National Safety Council.

1.11. NATIONAL SAFETY AWARD

National Safety Awards are given every year to give recognition for outstanding performance on the part of industrial establishments and to stimulate and maintain the interest of both managements and workers in accident prevention programmes. The award schemes were institute for factories registered under factories act, 1948, but from 1971, separate schemes for factories not covered under the act and for parts have also been introduced. The National Safety awards for mines were instituted in 1983.

1.12. GENERAL SAFETY RULES

In every organization a set of rules must be well formed to promote safety of employees and to prevent the direct and indirect costs of accident. Following points should be taken into consideration while formulating safety rules.

(1) The safety rules should be such that employees appreciate it and accept without any resistance.
(2) The safety rules should be such that it does not cause annoyance to an employee.
(3) The safety rules must be well defined.
(4) The safety rules should keep pace with changing environments and conditions.

Following points should be used as guideline to workers and supervisors.

(1) Smoking should be strictly prohibited, particularly near chemical or combustible materials.
(2) Personnel protective equipments like safety glasses, respirators, aprons, shoes and gloves must be always used depending upon type of work.
(3) Electrical connections and insulation should be checked at regular intervals.
(4) Electrical equipments must be operated by electric department personnel only.
(5) Safety guards must be provided on all rotating parts such as pulleys, gear boxes etc.
(6) Loose dress must never be used while working on machine.
(7) Material handling equipments should have unobstructed path for movement.
(8) Place the tools at its proper position.
(9) Mischievous acts should never be tolerated and defaulters should be punished.
(10) Prompt first aid attention must be paid to any insured person.

1.13. PERSONNEL PROTECTIVE EQUIPMENTS

Many companies employ personnel protective equipments as a means of accident prevention. An endless variety of personnel protective equipments are available. In general, personnel protective equipments can be classified as follows:

(*i*) Non-respiratory protective devices.
(*ii*) Respiratory protective devices.

Non-Respiratory Protective Devices

Personel protective devices for various parts of the body can be divided into five broad groups:

(*a*) Head Protection
(*b*) Eye and Face Protection
(*c*) Foot and Leg Protection
(*d*) Hand and Arm Protection
(*e*) Body Protection

(*a*) Hand Protection: The personnel protective equipments used for protecting the head include hard hats, helmets and caps. These are made of aluminium, PVC, fibre glass, laminated plastic or vulcanized fibre. They may be fitted with brackets for fixing welding masks, protective face screen, or a lamp. Helmets must be strong and sturdy enough to save workers from head injury specially due to falling objects, flying splinters, rotating machine parts etc. Soft caps and loads are also used for protection against heat, spark and other dangerous materials. These protective head wears should be fire resistant and where electrical hazards are present should be electrically insulated.

(*b*) **Eye-and-Face Protection:** Numerous eye injuries are caused by dusts, flying particles, splashes and harmful radiation. The devices used for eye protection may be safety spectacles, mono-goggles, impact goggles, welding goggles, foundry goggles, gas tight goggles, welding helmets, chemical goggles etc. The devices used for face and eye protection are face shield and the hood. The face shield provides protection to the entire face including eyes. Hoods are used under special circumstances for face and eye protection to deal with extremely hazardous situation as for example, working with highly corrosive substances. The face and eye protection devices should be heat resistant, scratch resistant, impact resistant and lenses should have antiglare properties.

(*c*) **Foot and Leg Protection:** Common foot and leg protective devices are safety shoes, boots, foot-guards, leg guards etc. Safety shoes may be meant for protection against for injuries, crushing of feet etc. Some other types of safety shoes which guard against electric shocks or harmful chemicals or inflammable etc., are also extensively used by workers. Boots usually of rubber are extensively used to provide protection against chemicals, slush etc. Gum boots are commonly used to work under wet conditions. The foot guard is a protective device meant to protect the toes up to ankles from injuries due to falling or striking objects. These are extremely used in foundries. Legging are wrapped around the legs up to knee or even up to hip.

(*d*) **Hand and Arm Protection:** Protection of hands and arm becomes necessary when workers have to handle materials having sharp ends, sharp edges or when hot and molten metals, chemicals corrosive substances have to be handled. The protective devices used may be gloves, mittens, hand guards and hand pads, sleeves, cuffs etc.

Gloves are used for complete protection of hands (including fingers) and are usually provided with wrist bands to ensure snug fit. They are made of rubber, leather, plastic, cotton etc.

Hand guards and hand pads are commonly used to save hands from possible injuries while working with trucks, gritty, sharp as well as highly heated objects.

Sleeves are meant for protection to arms (excluding hands) against injuries. Sleeves are made of leather, asbestos, plastic etc.

Cuff guards protect the lower part of the forearm from injuries specially when working in the proximity of heat sources, (for example, furnaces) including sparks.

The choice of material of these protective devices depends upon the tasks or operations to be performed.

(*e*) Body Protection: Sometimes it becomes necessary to provide special protective equipment for the body in the form of aprons, overalls, jackets and complete head to toe protective suits. The important considerations in the selection of safety clothing will depend upon the nature of potential hazard, degree of hazard involved and nature of activities of the user concerned. Although complete coverage of the body and legs is not needed in many cases and unnecessary safety clothing may hamper the efficiency of the user, no compromise should be made with strict safety requirements. If the user needs complete coverage, he should have it.

Some of the body protective clothing are as follows:

Apron: Aprons made of different materials are extensively used as protective clothing in various industries (welding, smithy, moulding) in particular. The selection of type of apron depends upon the nature of hazard arising out of radiation, acid, alkali and heat exposure etc. For example, lead rubber aprons are used for protection against radiation in welding operations.

Suits: Suits are specially designed for use in the extremely hazardous situation to provide complete body protection. Rubber suits are used for body protection in hazardous chemical plants white asbestos suits are suitable suits are suitable in case of fire hazards.

✎ Respiratory Protective Equipment

Atmospheric contaminants range from the relatively harmless substances to toxic dusts, fumes, smokes, mists, vapour and gases. To eliminate or minimize the hazard, the hazardous processes should, if possible, be closed or ventilated. If not possible or economical to close or ventilate such processes, respiratory equipments should be provided to the workers exposed to possible hazards.

Even though engineering means of control are applied satisfactorily, supply of appropriate protective equipment should be readily available for use. as there will be plant breakdown and repairs may have to be carried out in contaminated environments. Respiratory protective equipment should be considered as a last resort, or as standby protection and never a substitute for effective engineering control.

Types of Respiratory Protective Devices

CLASSIFICATION OF RESPIRATORY PROTECTIVE EQUIPMENT

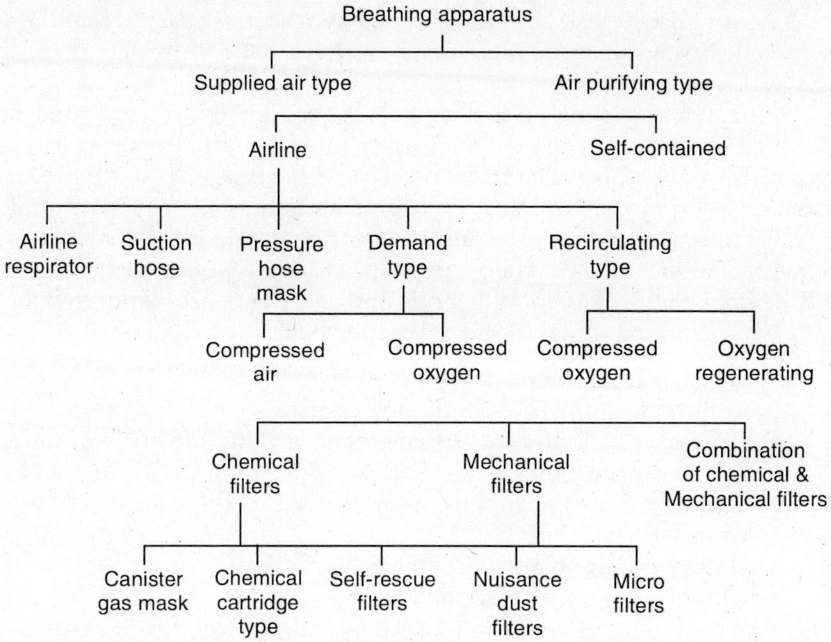

Airline Respirator: Airline respirator consists of a face-piece (half or full mask or a loose fitting helmet or hood) to which air is supplied through a small diameter hose. It may be a continuous flow type or a demand type. In a demand type air line respirator, the air from an air compressor is supplied to the face-piece through a demand valve which is actuated by the slight negative pressure created when the wearer inhales. On exhalation the demand valve closes and exhaled air escapes to the surrounding atmosphere through exhalation valve.

In a continuous flow type, air is supplied continuously to the face piece helmet or hood. Exhaled air or the excess air entering the face-piece escapes to the atmosphere.

Air supplied should be at least 110 liters of air per minute to enter the face piece and helmet or hood should be at least 170 liters per minute.

Airline respirators provide protection as long as the air supply is maintained but the wearer's travel is restricted by the length of the air supply hose.

Suction hose mark consists of a fuel piece connected to a large diameter flexible hose. The worker draws in air by his breathing effort. Air can be drawn in by breathing effort of the wearer up to 90 cm length of the hose.

Pressure hose mask is similar to suction hose mask except that the air is forced through a large diameter hose by a hand or motor operated blower.

Self-contained compressed air or oxygen breathing apparatus has compressed air or oxygen cylinder as its integral part. The air purifying type of respirators include (1) chemical cartridge type, canister type, hose mask type and mechanical filter type. The mechanical filter type respirators give protection against particulate air borne contaminants for *e.g.*, dust of cement, coal etc. The canister type respirators are also called gas mask are devices that cover mouth, nose and eyes and are connected to a canister containing suitable chemical absorbent.

1.14. HOUSE KEEPING

Good housekeeping is essential in order to:

- Make and maintain a clean, neat and orderly factory work areas and its surroundings.
- Make work area as look pleasant, more satisfying and motivating for a worker to work.
- Minimise fatigue and discomfort to workers.
- Minimise injury and accidents.
- Increase the life of plant, building and the facilities it contains.
- Avoid fire and other hazards.
- Permit effective natural illumination and ventilation.

➭ Procedure of Good Housekeeping

- Plan and protect the housekeeping programme carefully and completely. Associate the employees in the venture.
- Divide the plant and offices into cleaning zones and assign a person to each zone who will be responsible for good housekeeping and orderliness of his zone.
- Keep an eye on performed housekeeping schedule and conduct periodic houskeeping inspections. The following checklist may help in carrying out inspection properly.

(*A*) Machinery and Equipment

— General cleanliness
— Containers for waste materials
— Machine guards on and operating
— Oil, air, water, steam leakage
— Portable equipment. Do they hamper personnel and material movement?

(*B*) Materials and Storage:

— Piling and stacking – Can material slip easily?

— Materials protruding out of racks, bins, benches and machines etc.

(*C*) Building:

— Windows clean and unbroken

— Painting and upkeep

— Door joints clean

— Fire extinguishers and sprinklers clear

(*D*) Floors:

— Slippery, wet and oily

— Badly worn

— Garbage, dirt or debris

— Loose materials

(*E*) Stairways and Aisles:

— Clear and unblocked

— Well lighted

(*F*) Employee Facilities:

— Drinking taps clean

— Toilets and locker rooms clean

— Soap and towels available

(*G*) Other Aspects:

— Lamps and lamp reflector clean

— Bulletin boards and safety signs clean.

— Protective equipment and clothing clean and in good condition

— Electric motors clean

— Ventilation unobstructed

(*d*) Special rules for work done at a height in confined spaces, on certain electrical equipment or unguarded machinery.

(*e*) The maintenance of equipment and the provision of proper inspection and testing arrangements.

(*f*) General rules for safe working habits.

(*g*) Special rules for internal transport drivers.

(*h*) Arrangement for checking new machinery and materials.

(*i*) Safety inspection.

(*j*) The provision for personal protective equipment and rules as to its use.

(*k*) Suggestions on safety matters.

(*l*) Guidelines for environment protection.

IMPORTANT QUESTIONS

1. Define 'Safety'.
2. What is 'risk'?
3. What are the various measures of employee participation in safety?
4. Discuss the relevance of safety in the industry.
5. How plant productivity can be achieved through industrial safety? Explain.
6. What is plant layout? Discuss some of the safety considerations in plant layout.
7. What are different safety measures taken up in a organisation? Discuss.
8. How safety measures affect economy of an industry? Explain.
9. "Employees participation help of achieving safety objectives". Explain.
10. Explain relationship between safety and working environment.
11. Does equipment design affect safety? Explain.

❑❑❑

CHAPTER 2

Industrial Accidents

2.1. INDUSTRIAL ACCIDENT

An industrial accident may be defined as:

- *"An occurrence which interrupts or interferes with the orderly progress of work in an industrial establishment"*.

As per Factories Act 1948, industrial accident has been defined as:

- *"An occurrence in an industrial establishment causing bodily injury to a person which makes him unfit to resume his duties in the next 48 hours"*.
- *"An unexpected, unwanted event which cannot be anticipated in advance"*.

It is always a sudden process and not a gradual one. Moreover the event should be something to which a definite time, data and place can be assigned. It must occur in the course of employment in a factory or an industrial establishment.

2.2. NATURE OF ACCIDENTS

The nature of accidents may vary from industry to industry. An employee may be caught in a machine while working on it; or he may fall against a machine; or he may fall from a height while engaged on a particular task; or explosives used carelessly may explode. These accidents may result in disablement or death.

2.2.1. Disablement

Diablement whether **partial or total** may take the form of a loss of ability to work or to move. Such incapacity may be temporary or permanent.

2.2.2. Partial Disablement

Partial disablement means disablement of temporary nature that reduces the earning capacity of a workman in the employment in which he was engaged at the time of accident resulting in the disablement.

2.2.3. Temporary Partial Disablement

It means any disablement that temporarily reduces the earning capacity of the workman in every employment.

2.2.4. Permanent Partial Disablement

It means any disablement that permanently reduces the earning capacity of the workman in every employment.

2.2.5. Total Disablement

Total disablement means such disablement whether of a temporary or permanent nature, which incapacitates a workman for all work which he was capable of performing at the time of accident resulting in such disablement. Certain injures which are to be deemed to result in permanent total disablement. These include:

- Loss of both hands or amputation at higher sides
- Loss of hand and a foot
- Double amputation through leg or thigh or amputation through leg or thigh on one side and loss of other foot
- Loss of right to such an extent as to render the claiment unable to perform any work for which eye-sight is essential
- Very severe facial disfigurement, and
- Absolute deafness.

2.3. CAUSES OF ACCIDENTS

It is an established fact that accidents are caused, they do not just happen out of nothing. Whenever there occurs an accident, there must be some cause, which may be obvious, or difficult to trace. A large number of factors combine to cause accidents. Each of these factors is different in different situations. It has been rightly said that an accident does not have a single cause but a multiplicity of causes, which are often closely related. Basic factors responsible for causing accidents are *unsafe acts of the employees, unsafe working conditions and last but not least the chance occurrences.*

1. Unsafe Acts

Unsafe act may be defined as the deviation from the normal and correct procedure or practice. These acts may be result of lack of knowledge or skill on the part of the employee, certain bodily defects and wrong attitudes. Some of the examples of the unsafe acts are as follows:

(*i*) Failure on the part of workers to obtain necessary equipment or warning from other fellow workers of possible danger.

(*ii*) Operating or working at unsafe speeds, either too fast or too slow.

(*iii*) Operating machines or gadgets without approval of the relevant authority.

(*iv*) Throwing materials on the floor carelessly.

(*v*) Using unsafe equipment, or using equipment unsafely.

(*vi*) Failure on the part of workers to use safety devices and personal protective equipments.

(*vii*) Adopting unsafe or wrong methods in such operations like loading, unloading, removing, placing, intermixing etc.

(*viii*) Lifting goods especially fragile goods improperly.

(*ix*) Cleaning, adjusting, oiling, repairing, etc. or moving on a dangerous equipment.

(*x*) Distracting, tearing, abusing, quarrelling, day-dreaming, horseplay.

(*xi*) Taking unsafe positions under suspended loads.

Individual characteristics are also responsible for accident behaviour among workers. For example, characteristics like personality, motivation serve as a basis for certain behaviour tendencies such as tendencies to take risks and undesirable attitudes. These behavioural tendencies, in turn, result in unsafe acts as, failure to follow procedures and safety rules, and reluctance to use safety equipments, or haste on the part of the employee particularly, when he is on incentive payment schemes. These unsafe acts increase the probability of an individual incurring an accident.

2. Unsafe Conditions

Most of the accidents occur due to unsafe conditions *i.e.*, work-related causes. The work-related causes are often referred to as technical causes. These arise when there are improper or inadequate safety guards on machines; when machines break down; when improper personal protection equipment is installed; when mechanical or construction designs are defective and unsafe; and when control devices, which have been installed to make the operation of the machines safe and accident free are lacking or defective; or when there is an absence of proper maintenance and supervision of these devices.

Thus, unsafe conditions include:

(*i*) Improperly guarded equipment.

(*ii*) Defective equipment.

(*iii*) Hazardous arrangement of procedure in, and, or around, machines or equipment.

(*iv*) Unsafe storage; congestion; overloading.

(*v*) Inadequate safety devices.

(*vi*) Wrong and faulty lay out, and bad location.

(*vii*) Improper illumination — glare or insufficient light.

(*viii*) Improper ventilation — insufficient air charge, impure air source.

(*ix*) Poor housekeeping.

Safety experts are of the view that accidents can happen anywhere there are some "high danger" zones. For example, about one-third of industrial accidents occur around handlift trucks, wheel-barrows, and other handling and lifting areas. The most serious accident usually occur near metal and working machines and saws, or around transmission machinery like gears, pulleys, and flywheels, on stairs, ladders, walkways, scaffolds, hand rails. Handtools (like chisels or screw drivers) and electrical equipment (extension cords, electric drop lights, etc.) are other big accident causes.

The other work related causes of accidents are:

(*a*) *The job itself*— Some jobs are inherently more dangerous than others, such as the job of craneman in comparison to that of the foreman. Similarly, work in some departments (like personnel) is inherently safer than the work in others (like production department).

(*b*) *Work schedules*— Accidents increase late in the day. They do not usually occur during the early hours of the work day. They are more frequent during the night shifts. This is partly due to fatigue and partly due to the fact that night is the period when one requires rest.

(*c*) *Psychological climate of the work place*—It also affects the accident rate. Psychological, mental and emotional imbalance are at the root of several accidents. Emotionally disturbed and mentally pre-occupied persons meet with more accidents than a normal person. The psychological factors associated with accidents are fatigue, anxiety, tiredness, overwork, monotony, boredom, lack of self-confidence, and frustration. Fatigue often has a psychological origin, and may be due to anxiety, due to the monotony of the job on which a person is employed; due to frustration and inadequate incentives; due to unfair and incorrect methods of selection and promotion; due to the absence of group harmony and social integration; to bad leadership and ineffective organisation; and due to low social prestige. The frustrating situations includes sudden or arbitrary withdrawal of desirable objects and of certain privileges which were enjoyed by an employee; the creation of distrust in an individual and his ability to work; the creation of distrust on the part of employees of a particular, individual in the group; restrictions on opportunities for self-expression; discrepancy between an aspiration and the ability to solve a problem. All these situations affect for the worse the alertness of an employee, distract him, make him lose his concentration on the job in hand, and lead to accidents.

3. Other Causes

These causes of accidents arise mainly due to unsafe situational and climatic conditions. Some of these causes can be summarised as follows:

(*i*) Excessively long duration of work
(*ii*) Harsh and domineering behaviour of the management or supervisors towards workers.
(*iii*) Excessive heat in the factory premises.
(*iv*) Feeling of job insecurity among workers.
(*v*) Defective or reduced vision of the workers.
(*vi*) The nature of motivation adopted by management towards workers.
(*vii*) Reduced hearing capacity of the workers.
(*viii*) The leadership style adopted by the management in the organisation.
(*ix*) Lack of proper inspection and supervision by the supervisor.
(*x*) Careless handling of hazardous materials like petrol explosives etc.
(*xi*) Marital status of the workers. Unmarried employees generally have more accidents than married employees.
(*xii*) Addiction to alcohol and drugs.
(*xiii*) Smog in the atmosphere.
(*xiv*) Inexperienced workers generally sustain injuries than experienced workers.

2.4. CLASSIFICATION OF ACCIDENTS

There are several methods of classifying accidents some of the methods are as follows:

(1) Nature of Events
(2) Cause of Events
(3) Damage Caused
(4) Nature of Injury
(5) Length of Recovery.

(1) Nature of Events

According to the nature of events accidents are classified as:

(*a*) Traffic accidents
(*b*) Passenger accidents

(*a*) Traffic Accidents:

(*i*) Collisions with cars, buses and motor vehicles.
(*ii*) Collisions with pedestrians.

(*b*) Passenger Accidents:

(*i*) Slips, stumbles, falls, etc.
(*ii*) Boarding moving car or bus.
(*iii*) Strucking by doors.

(2) Cause of Events

According to the cause of events accident are classified as:

(*a*) Machine accidents.
(*b*) Non-machine accidents.

(*a*) Machine Accidents:

(*i*) Catching of tools, guides, flying particles in machines.
(*ii*) Catching of fingers, arms, legs, clothing etc. in machines.

(*b*) Non-machine Accidents:

(*i*) Trailing of telephone cables on floor.
(*ii*) Objects on floor.
(*iii*) Pushes, bumps etc. by other persons, objects.

(3) Damage Caused

(*a*) Damage of property
(*b*) Damage of material
(*c*) Damage of building.

(4) Natural of Injury

(*a*) Fatal accidents
(*b*) Permanent accidents
(*c*) Temporary accidents.

(*a*) Fatal Accident: In such a accident persons are killed.

(*b*) Permanent Accident: Due to this accident employee loses earning capacity. The employee is not capable of earning that much money which he was capable to earn before accident. The employee may cut his finger, arm or hand. It is of two types (*i*) Total disablement (*ii*) Partial disablement.

(*c*) Temporary Accident: In this accident worker is unfit temporarily and loses his earning capacity for a short time. Fracture of arm is example of temporary accident.

(5) Length of Recovery

According to length of recovery accidents are classified as:

(*a*) First Aid cases

(*b*) Home case

(*c*) Lost time

(*a*) First Aid Cases: In these cases injuries are minor. The workers are given first aid in the factory. After treatment the workers are ready to work. In this case no time is lost except when the first aid treatment is given.

(*b*) Home Case: In this case medical treatment is firstly given at factory hospital and then worker is allowed to go home. The worker is allowed to be away from duties for a maximum of 3 days. The worker recovers in this period and then resumes his duty. No compensation is paid in this case.

(*c*) Lost Time Accidents: In this case the injuries are serious and workers are admitted to hospital. The worker is away from duty from more days and compensation is paid to the workers. The investigation of accidents is done to find the causes of accidents.

2.5. ACCIDENT-PRONENESS

It is condition in which a "*human being is mentally inclined, strongly disposed, attitudinally addicted or personally destined to become continually involved in an ongoing and never-ending series of accidents or injuries*". Accordingly some people may be more often involved in accidents than others. In other words, they are "*accidents-prone*". They get involved in accidents, however hard they try to avoid them. The causes of accident-proneness are muscular weakness, emotional instability, visual disability, recklessness, hostility and indifference.

The concept of accident-proneness, however, is now being replaced by that of "accident repeater", that is an accident repeater generally has more than a normal share of accidents during a particular period of time. Repeated accidents are not due to any mysterious or supernatural cause but are due to the personality, emotional make-up, family background and physical condition of an individual in the context of specific circumstances. Age, experience and supervision are also sometimes related to the recurring nature of accidents and injuries.

2.6. TYPES OF ACCIDENTS

Depending on the severity of the injury, accidents may be typed as:

- Major accident
- Minor accident

An accident which results in a death or prolonged disability to the victim is a **major** one. A scratch or cut which does not seriously disable one is a **minor** accident. An accident may be **internal** or **external**. If a worker falls or an object falls on him, it may not show signs of injury but they may have fractured a bone or strained a muscle or nerve which is an **internal** injury. A deep scratch on the leg or shoulder may show sign of injury which is an **external** injury. A worker may be disabled by injury for a day or week or month or for a few months. If he recovers from such a disability, his disability is **temporary**, but if the injury is such that he will never recover fully, his disability is **permanent** one.

A disability may be **partial** or **total**. An accident may be **fatal** or **non-fatal**. Figure 2.1 shows the various types of accidents.

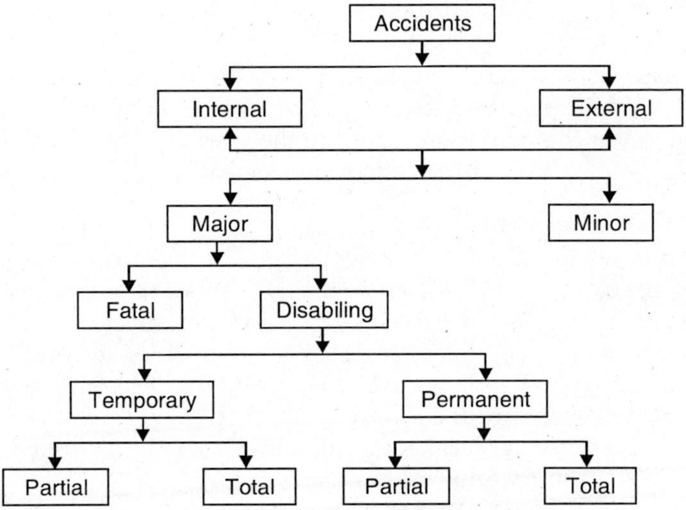

Fig. 2.1. Types of Accident

2.7. ACCIDENT COSTS

Accident cause enormous crisis directly or indirectly. The losses incurred due to accidents could be visible as well as invisible. Certain losses can be evaluated in terms of money whereas certain other losses cannot be evaluated in monetary terms. Whenever an industrial accident occurs, it gives rise to pain for the victim and his family and retards industrial productivity which in turn, affect the very economic base of the industry. In addition to this, compensation to be paid to the employees involved in an accident is enormous.

The management suffers losses involving (*i*) *Direct costs* and (*ii*) *Indirect costs.*

(*i*) **Direct costs**, *i.e.*, the wages of employees is six to ten times the normal wages because of the loss of goods and services, compensation and the cost of medical aid, the cost incurred on training a new worker, loss due to waste of raw materials; and loss of production and quality arising out of the inexperience and lack of skill of the new employee.

(*ii*) **Indirect costs**, which include the following:

(*a*) The cost which the government has to incur because it has to maintain a larger number of factory inspectors to check accidents and it has to spend more on the employee's health insurance and other social security benefits and the cost of all these is recovered by imposing higher taxes on the people.

(*b*) The cost of the employee time because he has been without work because of his accident.

(*c*) The cost of the lost time because other employee stop work, out of curiosity, out of sympathy with the injured employee or because they have to assist the injured worker.

(*d*) The cost of time lost by a foreman, a supervisor or other executive while assisting the injured employee, investigating the cause of the accident, arranging for his replacement, selecting and training a new employee, preparing the accident report, and attending hearings conducted by government or other officials;

(*e*) The cost incurred on the machine or tools that might have been damaged, and/or the cost of the spoilage of material when the accident occurred.

(*f*) The loss of profits on the production which the injured employee would have been responsible for, including the loss incurred because the machine on which he was working was idle;

(*g*) The cost incurred on account of the wages paid to an employee during the period in which he was idle following his injury and even after his return to work, when his production would be worth much less than it was before he sustained the injury;

(*h*) The loss due to the weakened morale of the other employees following the occurrence of the accident and the consequent lower productivity throughout the plant and

(*i*) Overhead costs: the expense incurred on light, heat, rent and such other items, which continue to be used while the injured employee is a non-producer.

It is obvious then, that an accident causes a lot of suffering and loss to the employees, the employer, the government, and even to society. It is particularly hard on the employee's family, specially when he is the only bread-winner. Every accident lowers the morale of his fellow-workers. They become pessimistic and increasingly aware of hazardous nature of their work, as a result of which they do not, or cannot, put their best efforts to achieve optimum production. Accidents therefore, increase the overall cost of production, and adversely affect productivity and morale.

2.8. WELFARE AND SAFETY

Welfare work aims at providing such service facilities and amenities as would enable the workers employed in industries/factories to enable their work in healthy congenial surroundings conducive to good health and a high morale. It is partly

(*a*) humanistic because it enables the workers to enjoy a fuller and richer life.

(*b*) economic because it improves the efficiency of workers, increases its availability where it is scarce and keeps them contented. It therefore, minimises the inducement to form or join unions and to resort to strike.

(*c*) civic because it develops a sense of responsibility and dignity among the workers and thus makes them worthy citizens of the nation.

Safety is basically concerned with minimising hazards and occupational diseases in industries. In certain cases welfare and safety activities are closely linked together. These areas may include the following:

(*a*) Proper layout of plant and machinery.

(*b*) Provision of medical facilities.

(*c*) Provision of first aid appliances.

(*d*) Provision of compensation in case of accidents.

(*e*) Proper ventilation.

(*f*) Provision of fire extinguishers.

(*g*) Provision of personnel protective devices.

(*h*) Fencing and guarding of machines.

(*i*) Improving working conditions.

(*j*) Reasonable hours of work.

(*k*) Temperature and humidity control in the workplace.

The areas in which safety and welfare are not directly linked are:

(*a*) Provision of rest place, canteens, restaurant and cafeteria.

(*b*) Provision of kindergartens and nurseries.

(*c*) Financial assistance example provident fund, gratuity, pension etc.

(*d*) Housing facilities.
(*e*) Educational facilities.
(*f*) Recreational facilities.

2.9. ACCIDENT MEASUREMENT

Two main statistical ratios have been used to gather accident information—*the frequency rate* and *severity rate*.

The frequency rate is the number of time-lost accidents (or injuries which have disabled an employee) per 1,000,000 manhours worked. The severity rate, on the other hand, is the total number of days charged or lost because of accidents per 1,000,00 manhours worked.

The National Safety Council of the United States has given the following formulae for the computation of these rates:

Accident Frequency Rate =
Number of disabling work injuries × 1,000,000/
Total number of manhours worked

Severity Rate = Number of mandays lost × 1,000,000/
Total number of manhours worked

NUMERICAL EXAMPLES

Example 2.1. Using the following data calculate the frequency rate of accident in an industrial plant.
(*a*) Number of workers,
(*b*) Number of disabling injuries per year = 5
(*c*) Average number of hours worked by worker per year = 2000

Solution: $$\text{Frequency Rate} = \frac{\text{Number of disabling injuries}}{\text{Number of manhour worked}} \times 1000{,}000$$

$$= \frac{5}{1000 \times 2000} \times 1000{,}000 = 2.5.$$

Example 2.2. Using following data calculate reverity rate of accident in an industrial plant where only one accident occurred during the day involving
(*a*) Cutting the tip of a finger.
(*b*) Total disability of worker
The number of workers = 2000, Number of days lost in a year due to accident = 100, Average number of hours worked by worker per year = 2000.

Solution: $$\text{Severity Rate} = \frac{\text{Days lost in a year due to accident}}{\text{Number of manhour worked}} \times 1000{,}000$$

Days lost =Actual days lost due to accidents + standard number standard number is 35 days if nature of injury is being cutting of tip of fingers whereas standard number is 6000 days for a total disability case.

(*a*) Days lost = 100 + 35 = 135 days

$$\text{Severity Rate} = \frac{135}{2000 \times 2000} \times 1000{,}000 = \frac{135}{4} = 33.75$$

(*b*) Days lost = 100 + 6000 = 6100 days

$$\text{Severity Rate} = \frac{6100}{2000 \times 2000} \times 1000{,}000 = \frac{6100}{4} = 1525.$$

CASE STUDIES

Developing countries such as India, are particularly vulnerable to industrial crises as they lack the infrastructure (*e.g.*, communication, training education etc.) required to maintain technology but are nevertheless, eager to set up and maintain industrial plants. As a result, they compete globally to attract multinational companies for their investment and capitals, and in this process, often tend to ignore the safety and health violations that many MNC's engage in. "Developing countries confer upon MNC's a competitive advantage because they offer low-cost labour, access to markets, and lower operating costs. Once there, companies have little incentive to minimize environmental and human risks. Lax environmental and safety regulation, inadequate capital investment in safety equipment, and poor communications between companies and governments compound the problem."

The Bhopal facility was part of India's Green Revolution aimed to increase the productivity of crops. Considered an essential factor in the effort to achieve self-sufficiency in agricultural production pesticide production use increased dramatically during the late 1960's and early 1970's. The decision to manufacture the pesticides in India, as opposed to relying on imports was based on India's goal of preserving foreign exchange and its policy of industrialization. In 1969, Union Carbide (UCC-the parent company) set up a small plant (Union Carbide India Ltd.-UCIL) in Bhopal, the capital city of Madhya Pradesh, to formulate pesticides.

Bhopal was chosen as the site of the Carbide plant because of it's central location in India, a railway system that spanned that country, a large lake which provided a reliable source of water, and sufficient electricity and labour to sustain a large scale industrial plant. The MIC facility was located in the existing. Carbide plant to the north of the city, adjacent to an existing residential neighbourhood and barely two kilometers from the railway station. Union Carbide claims that the "squatter settlements" around the plant arrived only after it did. However, "all three of the worst-affected communities in the disaster apparently existed before the Union Carbide plant opened".

Until 1979, the Indian subsidiary of Carbide used to import MIC or methyl isocyanate from the parent company. After 1979, it started to manufacture its own MIC. MIC is one of many "intermediates" used in pesticide production and is a dangerous chemical. It is a little lighter than water but twice as heavy as air, meaning that when it escapes into the atmosphere it remains close to the ground. It has the ability to react with many substances: water acids, metals, and the small deposits of corrosive materials that accumulate in pipes, tanks, and valves. The MIC in Bhopal was used for the production of various pesticides, mainly Sevin brand carbaryl insecticide and Temik brand aldicarb pesticide. All the pesticides produced at UCIL were sold in the Indian market.

According to many, Bhopal is the site of the greatest industrial disaster in history. On the night of December 23, 1984, a dangerous chemical reaction occurred in the Union Carbide factory when a large amount of water got into the MIC storage tank # 610. The leak was first detected by workers about 11: 30 p.m. when their eyes began to tear and burn. They informed their supervisor who failed to take action until it was too late. In that time, a large amount, about 40 tons of Methyl Isocyanate (MIC), poured out of the tank for nearly two hours and escaped into the air, spreading within eight kilometers downwind, over the city of nearly 900,000. Thousands of people were killed (estimates ranging as high as 4,000) in their sleep or as they fled in terror, and hundreds of thousands remain injured or affected (estimates range as high as 400,000) to this day. The most seriously affected areas were the densely populated shanty towns immediately surrounding the plant — Jayaprakash Nagar, Kazi Camp, Chola Kenchi, and the Railway Colony, the victims were almost entirely the poorest members of the population.

This poisonous gas, caused death and left the survivors with lingering disability and diseases. Not much is known about the future medical damage of MIC, but according to an international medical commission, the victims suffer from serious health problems that are being misdiagnosed or ignored by local doctors.

Exposure to MIC has resulted in damage to the eyes and lungs and has caused respiratory ailments such as chronic bronchitis and emphysema, gastrointestinal problems like hyperacidity and chronic gastritis, ophthalmic problems like chronic conjunctivitis and early cataracts, vision problems, neurological disorders such as memory and motor skills, psychiatric problems of various types including varying grades of anxiety and depression, musculoskeletal problems and gynecological problems among the victims. It is estimated that children born in Bhopal after the disaster face twice the risk of dying as do children elsewhere, partly because parents cannot care for them adequately. Surprisingly enough, despite the serious health problems and the deaths that have occurred, Union Carbide claims that the MIC is merely a "mild throat and ear irritant".

The Bhopal disaster was the result of a combination of legal, technological, organizational, and human errors. The immediate cause of the chemical reaction was the seepage of water (500 liters) into the MIC storage tank. The results of this reaction were exacerbated by the failure of containment and safety measures and by a complete absence of community information and emergency procedures. The long term effects were made worse by the absence of systems to care for and compensate the victims.

Furthermore, safety standards and maintenance procedures at the plant had been deteriorating and ignored for months. A listing of the defects of the MIC unit runs as follows:

— Gauges measuring temperature and pressure in the various parts of the unit, including the crucial MIC storage tanks, were so notoriously unreliable that workers ignored early signs of trouble.
— The refrigeration unit for keeping MIC at low temperatures (and therefore less likely to undergo overheating and expansion should a contaminant enter the tank) had been shunt off for some time.
— The gas scrubber, designed to neutralize any escaping MIC, had been shut off for maintenance. Even had it been operative, post-disaster inquiries revealed, the maximum pressure it could handle was only one-quarter that which was actually reached in the accident.
— The flare tower, designed to burn off MIC escaping from the scrubber, was also turned off, waiting for replacement of a corroded piece of pipe. The tower, however, was inadequately designed for its task, as it was capable of handling only a quarter of the volume of gas released.
— The water curtain, designed to neutralize any remaining gas, was too short to reach the top of the flare tower, from where the MIC was billowing.
— The lack of effective warning systems; the alarm on the storage tank failed to signal the increase in temperature on the night of the disaster.
— MIC storage tank number 610 was filled beyond recommended capacity; and

— A storage tank which was supposed to be held in reserve for excess MIC already contained the MIC.

Ironically, in Bhopal, the people living around the Union Carbide plant were warned of potential hazards in a series of local newspaper articles, but residents ignored these warnings because they did not know how to react to them, while local official dismissed them as sensationalist reporting.

Interestingly enough, Carbide tried to hide its poor safety and maintenance record along with the other faults already mentioned, by claiming publicly that the company was the victim of sabotage by a 'disgruntled employee'. Yet, Carbide didn't release the name of this employee or bring charges against him/her. However, there is evidence to the contrary which supports the view that Carbide (both the parent company and its Indian subsidiary) was a negligent company that failed to improve its deteriorating plant. Incidentally, in a report (May 1982) of the Indian subsidiary conducted by a three-member safety team from the Union Carbide headquarters in the U.S., indicated that "a serious potential for sizeable releases of toxic materials in the MIC unit either due to equipment failure, operating problems, or maintenance problems thus requiring various changes to reduce the danger of the plant; there is no evidence the recommendations were ever implemented".

Furthermore, "Carbide persistently shows wanton and wilful disregard for the health and safety of its workers and the communities in which it operates.'" Additionally, a scientific report published by two U.S. organizations, the National Toxic Campaign and the international Council on Public Affairs, Union Carbide continues to be "a major discharger of toxic substances into the environment, and a major generator of hazardous waste'. In 1988, the company generated more than 300 million pounds of hazardous waste – an increase of 70 million compared with 1987".

Carbide had dropped the safety standards at the Bhopal plant well below those it maintained at a nearly identical facility in West Virginia It is also important to note here the Carbide was able to operate its deteriorating plant because industrial safety and environmental laws and regulations were lacking or were not strictly enforced by the state of Madhya Pradesh or the Indian government making them indirectly responsible for the tragedy at Bhopal.

ASSESSMENT OF AIRBORNE DUST ASSOCIATED WITH CHEMICAL PLANT: A CASE STUDY

Alumina refinery is a mineral-based chemical plant involving refining of crushed bauxite ore using various chemicals and minerals as input materials like caustic soda, quicklime, coal, wheat bran and mineral acids.

The process of alumina production involves refining of bauxite ore into tri-hydrated alumina (A1203, $3H_2O$) by chemical method followed by process of calcinations. This method possesses, various kinds of dust hazards in its work environment amongst the people involved. Poor health of industrial employees in India is due to its occupational environment which is a major concern nowadays. Attempts have been made to recognize the potential sources of airborne dust and to assess the dust load upon exposed workers at different work sites of alumina plants by comparing the observators with the standard values called 'Threshold Limit Values' (T.L.V.) assigned by the international body ACGIH (American Conference of Governmental and Industrial Hygienists), USA and also permissible exposure limit values prescribed in the second schedule Section F of Factories Act (Amendment), 1987. Alumina plant operation includes various physical operations like crushing, grinding, conveying, loading, transporting, etc., which generate finer particles. It can cause serious health hazards on inhalation, depending upon its size, shape, constituents and duration of exposure. Out of all these parameters, concentration of respirate fraction of airborne dust (0.5 to 5.0 micron size) and its free silica content have been reported to cause lung fibrosis as well as occupational disorders. In the present study, attempts have been made to make a survey of respirable fraction of the airborne dust (that remains suspended in air for quite an appreciable time) associated with various operations according to job profiles. It also outlines the probable control measure in order to provide a healthy working environment. Present work aims at identifying and evaluating the degree of workplace dust with special reference to respirable fraction and for recommending suitable suggestive control measures for an effective management of occupational environment.

2.10. ACCIDENT PREVENTION

There are several ways by which accidents can be prevented. The**National Safety Council U.S.A.**, says that accident prevention depends on the three E's - *Engineering, education and enforcement*: the job should be engineered for Safety; employees should be educated in safe procedures; and safety rules should be properly enforced.

In actual practice the prevention of accidents may be effected through the following measures

(*i*) Maintenance of a proper working environment *e.g.*, proper distribution of light, glare control, noise control, proper ventilation etc.

(*ii*) By proper repair and maintenance of the factories, other premises, structures, plants and equipments etc.

(*iii*) By proper planning and designing including structural designing at the grass root level.

(*iv*) By proper selection and placement of employees, safety training to new employees in safe practices and procedures and working of potential hazards; and persvasion and propaganda.

(*v*) Taking correct safety measures by workers and the management through safety consciousness programmes *e.g.*, safety weeks, safety slogans, safety campaigns, safety quizes etc.

(*vi*) By proper supervision, checking and inspection of the various processes relating to industrial complexes or factories.

(*viii*) Proper checking of the materials so that sub-standard materials are weeded cut.

2.11. INVESTIGATION AND ANALYSIS OF ACCIDENTS

The accident prevention programme involves quick, positive and determined efforts to pinpoint the unsafe acts and unsafe conditions and to adopt the correct measures. Inspite of the efforts the accidents occur, hence the investigation of the accidents become one of the important function in the accident prevention programme. The objective of the investigation is to determine the basic causes of accident in order to take remedial action to prevent recurrence and thus reinforce the safety programme activities. In the investigation of accidents care should be taken that:

(*i*) *The investigation should be carried out by supervisor who is well conversant with the situation.*

(*ii*) *In order to obtain true picture, the investigation should be carried out without any delay because true cause cannot be detected due to possibility of change of situation.*

(*iii*) *The investigation should be honest and should have the aim of exploring the facts and not to blame or fix up the responsibility on somebody.*

The supervisor who is assigned the task of investigation of accidents will prepare the report of the investigation and recommendations in an appropriate form and submit to the concerned higher authority.

The investigation report is further analysed in details by proper data compilation to know the areas of attack for accident prevention.

The accident report form may vary with specific cases, still the report form should include:

(*i*) *The facts or identification.*

(*ii*) *A narrative description of the accident.*

(*iii*) *The major cause of accident, namely, the unguarded machine or other mechanical and physical hazard, the unsafe act of person.*

(*iv*) *Facts of prevention namely, what was done and what should be done to prevent the occurrence of other similar accidents. A typical sample form is as follows.*

Supervisor's Report of Accident Investigation Identification

Company ______________ Deptt______________ Location ____________
Time of Accident ___________ Date of investigation ________________
Name of injured _________________Age __________Position __________
Severity of injury ———————— Treatment ————————Doctor __________ Witness of accident ____________.

(*a*) How did it happen?

(*b*) Which machine, tool, substance, objects was most closely connected with accident?

(*c*) If a machine or vehicle, what part of it?

(*d*) In what way was the machine, tool, object are substance unsafe?

(*e*) Were mechanical guards or other safeguards provided?

(*f*) Was the injured person using them when accident occurred?

(*g*) Name the unsafe practices of the injured person or other person that resulted in this accident.

(*h*) Why did the person act unsafely?

(*i*) How can you prevent accidents of this type?

(*j*) What have you already done to prevent a similar accident?

Signature of safety Director **Signature of Supervisor**

✎ IMPORTANT QUESTIONS

1. Define industrial accident.
2. What are the various types of accidents in an industry.
3. What are the factors responsible for industrial accidents.
4. Explain 'Accident-Proneness'.
5. What are the types of accident costs?
6. How accident cost can be measured?
7. What are different measures taken up in a organization to prevent accident.

❒❒❒

CHAPTER 3

Industrial Hazards

3.1. INTRODUCTION

Industrial Hazard refers to a *potential condition which might be converted into an event (accident)*. It is a state, physical or chemical having potential to injure the person or impairment of health. Risk or danger arises out of hazards. Hazards may be broadly grouped under the following head:

(*i*) Chemical Hazards
(*ii*) Mechanical Hazards
(*iii*) Biological Hazards
(*iv*) Environment or physical Hazards
(*v*) Electrical Hazards
(*vi*) Fire Hazards
(*vii*) Potential Hazards

(*i*) Chemical Hazards

Chemical substances, such as carbonmonoxide, carbon dioxide, nitrogen oxide, sulphur dioxide, hydro-carbons, ozone, sulphuric acid, acetic acid, fumeric acid and tannic acid, limes and alkalies cause injury when they are absorbed by the skin, or when they are ingested or inhaled. The result are often disastrous. Workers may suffer from respiratory diseases, skin diseases, allergy, heart disease, cancer and neurological disorder, all of which often shorten life expectancy. The disease of sickness may be chronic or acute, and it may appear after a long dormant period when it is difficult or impossible to treat it effectively. Often, a disease may be difficult to diagnose because its symptoms are not apparent at all.

Gases, fumes and dust raised by such processes as grinding and crushing of stones or minerals may be inhaled by workers and cause a serious injury, or even death coalminers often suffer from what is known as "*black lung*" disease. Employees in manufacturing industries are often exposed to such health hazards which arise from dust and fumes, while those working on lead or zincs melters often show indications of zinc or lead poisoning.

(*ii*) Mechanical Hazards

Mechanical hazards can mainly stem from any of the following causes:

(*i*) *Failure of components of a mechanical system due to faulty design, faulty assembly, faulty materials, lack of repair and maintenance.*

(*ii*) *Inadvertance and complacency on the part of workers at the points of operation.*

(*iii*) *Inadequate safeguards on the machine.*

(*iv*) *Laxity in the use of personal protective equipments.*

The chief sources of mechanical hazards originate from flywheels, protruding parts of a moving machinery, presses pressure vessels, all sorts of drives namely belt, chain and sprocket; conveyers, lifts, hoist tackles, shafts, hand tools, power tools etc. The operations which make the workers vulnerable to injuries are chiefly cutting, shearing, breaking, tearing, crushing, hoist, pressing etc. As a consequence of these operations, the workers can injure themselves in a number of ways like cutting of finger tips, crushing of limbs, injury due to flying splinters, fall from great heights, tangling by rapidly rotating machine parts.

(*iii*) Biological Hazards: These include bacteria, viruses, fungi, and other living organisms that can cause acute and chronic infections by entering the body either directly or through breaks in the skin. Occupations that deal with plants or animals or their product or with food and food processing may expose workers to biological **hazards**. Laboratory and medical personnel also can be exposed to biological **hazards**. Any occupational that result in contact with bodily fluids pose a risk to workers form biological **hazards**. In occupations where animals are involved, biological **hazards** are dealt with by preventing and **controlling** diseases in the animal population as well as properly caring for and handling infected animals. Also, effectively personal hygiene, particularly proper attention to minor cuts and scratches especially on the hands and forearms, helps keep worker risks to a minimum.

In occupations where there is potential exposure to biological **hazards** workers should practice proper personal hygiene, particularly hand washing. Hospitals should provide proper ventilation, proper personal protective equipment such as gloves and respirators, adequate infectious waste disposal systems, and appropriate controls including isolation in instances of particularly contagious disease such as tuberculosis.

***(iv)* Environment or Physical Hazards:** These include excessive levels of ionizing and nonionizing electromagnetic radiation, noise, vibration, illumination and temperature.

In occupations where there is exposure to ionizing radiation, **time, distance,** and **shielding** are important tools in ensuring worker safety. Danger from radiation increases with the amount of time one is exposed to it; hence, the shorter the time of exposure the smaller the radiation danger.

Distance also is a valuable tool in **controlling** exposure to both ionizing and non-ionising radiation, radiation levels from some sources can be estimated by comparing the squares of the distances between the worker and the source. For example, at a reference point of 10 feet from a source, the radiation is 1/100 of the intensity at 1 foot from the source.

Shielding also is a way to protect against radiation. The greater the protective mass between a radioactive source and the worker, the lower the radiation exposure. In some instance, however, limiting exposure to or increasing distance from certain forms of non-ionizing radiation, such as lasers is not effective. For example, an exposure to laser radiation that is faster than the blinking of an eye can be hazardous and would require workers to be miles from the laser source before being adequately protected. Shielding workers form the source can be an effective control method. Noise, another significant **physical** hazard, can be controlled by various measures. Noise can be reduced by installing equipment and systems that have been engineered, designed, and built to operate quietly; by enclosing or shielding noisy equipment; by making certain that equipment is in good repair and properly maintained with all worn or unbalanced parts replaced; by mounting noisy equipment on special mounts to reduce vibration; and by installing silencers, mufflers, or baffles.

Substituting quiet work **methods** for noisy ones is another significant way to reduce noise—for example, welding parts rather than riveting them. Also, treating floors, ceilings, and walls with acoustical material can reduce reflected or reverberant noise. In addition, erecting sound barriers at adjacent work stations around noisy operations will reduce worker exposure to noise generated at adjacent work stations.

It is also possible to reduce noise exposure by increasing the distance between the source and the receiver, by isolating workers in acoustical booths, limiting workers' exposure time to noise, and by providing hearing protection.

(v) Electric Hazards

Necessary protection and safety devices are warranted to contain electric hazards. Electric exposures can cause a myriad of problems which must be judiciously countered. The protection programme should envisage safety to:

(a) Users or occupants or workers.

(b) Electrical machines and equipments.

(c) The electrical system encompassing conductors, switches etc.

(d) Fire-hazards associated with short circuits etc.

Under normal circumstances the rate of current flow in amperes is of the order of 20 mA to 25 mA can cause electrical shocks. A fatal shock can be caused by 50 mA of current. All the above relate to dry skin conditions. In case of wet skin condition, of human body, the resistance could be even 100th part compared to the dry skin conditions.

Hence all occupants or workers should be warned not to touch any electrical installation and machines including switches in the wet skin conditions to avoid fatal shocks. A slight delay can cause fatal accidents or inflaming of the electric system.

(*i*) For household wiring connections and small units, a fuse can be provided as a safety measure. The fuse interrupts the electric supply in case overloading or faults. It may be better to use miniature circuit breakers (MCBs).

(*ii*) For big industrial units (ACBs) air circuit breaker as well as (OCBs) oil circuit breakers should be used to protect electric installations from hazards. Moulded Cases Circuit Breakers (MCCBs) or Earth Leakage Circuit Breakers (ELCBs) give superior performance. These circuit breakers can cut-off the supply immediately if the current exceeds a particular stipulated limit within a fraction of seconds, they can as well cut-off the main supply during short circuits or leakages.

(*iii*) For all electric connections proper earthing or grounding should be done.

(*iv*) All live electric machines, installations or gadgets which are liable to cause shocks should be properly guarded by enclosures or railings with danger rings signs properly displayed.

(*v*) All electrical apparatus in industries must be enclosed inside flame proof boxes or enclosures to avoid the danger which may result from sparking during the operation of switches.

(*vi*) Provision of lighting conductors should be made to protect the workers as well as the electrical installations or electrical equipment from lighting.

(*vii*) Safety norms as mentioned in the Indian Electricity Rule, 1956 along with the amendments must be strictly followed.

(*viii*) Inspite of best care and protection by using earthing and insulation, the control method may fail. Under such circumstances use of an improved device called Ground fault Interrupter (GFI) is suggested. It automatically interrupts the circuit practically instantly.

(*ix*) All circulation spaces and working spaces in the vicnity of hazardous spots should adequate.

(*x*) Use of rubber gloves and mats should be specially recommended for workers working in live wires, high installations and where bare conductor carries current of high amperage.

(*vi*) Fire Hazards

The hazards involved in the construction processes and their fall-out could be summarised as follows:

(*a*) *Those hazards or accidents that can cause immediate death.*

(*b*) *Those hazards or accidents which can cause injury or loss of limbs of the workers.*

(*c*) *Those hazards or accidents which can cause only loss of property.*

These above three categories will have instantaneous or immediate ill-effects.

Factors responsible for causing hazards or accidents are as follows:

(*i*) Faulty structural design.

(*ii*) Improper repair and maintenance.

(*iii*) Substandard and defective materials of construction which do not satisfy the relevant standards and specifications.

(*iv*) Improper planning of layout.

(*v*) Lack of supervision during construction.

(*vi*) Ignoring the norms of sound engineering practice.

(*vii*) Carelessness and irresponsibility of the workforce in the execution of construction.

The following precautions may be adopted to minimise hazards due to the above factors.

(*i*) Structural design must be entrusted to qualified and experienced structural engineers.

(*ii*) By carrying out regular repairs and maintenance under the expert guidance of qualified and experienced engineers.

(*iii*) Materials of construction must be tested for strength and durability by standard laboratories.

(*iv*) The planning of the structures etc. should be entrusted to architects or engineers having sound knowledge of safety principles and practices.

(*v*) All supervision work must be done under the expert supervision and guidance of qualified and experienced engineers or architects.

(*vi*) Care must be taken that the norms of sound engineering practice are not violated.

(*vii*) The workforce responsible for construction must be instructed to remain careful and responsible and follow safety rules and norms.

(*vii*) Potential Hazards:

It include dusts, fumes, mists, aerosols, fibres and gases

Dusts are solid particles generated by handling, crushing, grinding, colliding, exploding and heating organic or inorganic materials such as rock, ore, metal, coal, wood and grain.

Fumes are formed when material from a volatilized solid condenses in cool air. In most cases, the solid particles resulting from the condensation react with air to form an oxide.

The term **mist** is applied to liquid suspended in the atmosphere. Mists are generated by liquids condensing from a vapour back to a liquid or by a liquid being dispersed by splashing or atomizing. **Aerosols** are also a form of a mist characterized by highly respirable, minute liquid particles.

Fibers are solid particles whose length is several times greater than their diameter, such as asbestos.

Gases are formless fluids that expand to occupy the space or enclosure in which they are confined. They are atomic, diatomic, or molecular in nature as opposed to droplets or particles which are made up of millions of atoms or molecules. Through evaporation, liquids change into vapours and mix with the surrounding atmosphere. **Vapours** are the volatile form of substances that are normally in a solid or liquid state at room temperature and pressure. Vapours are gases in that true vapors are atomic or molecular in nature.

3.2. HAZARD IDENTIFICATION TECHNIQUES

In the 1960's larger chemical plants resulted in larger accidents. The traditional approach to safety management was to have an accident, investigate and then make changes to prevent a repetition. This approach is no longer acceptable. A single incident can challenge the existence of a company due to the damage to reputation, loss of assets and production. Hazard identification techniques were developed to anticipate problems before they occur. Hazard identification techniques start from the principle that *all work related injuries and illnesses and escapes of hazardous material can be prevented.* A number of different methods can be used for identifying hazards. The most widely used of these are as follows:

(*i*) *Failure mode and effects analysis (FMEA) Technique*
(*ii*) *Hazard and operability review (HAZOP) Technique*
(*iii*) *Technique of operations review (TOR)*
(*iv*) *Human Error analysis (HEA) Technique*
(*v*) *Fault Free analysis (FTA) Technique*
(*vi*) *Risk analysis Technique*

➥ Failure Mode and Effects Analysis (FMEA) Technique

FMEA is formal, step-by-step analytical method that is a spin-off of reliability analysis, a method used to analyse complex engineering systems. FMEA proceeds as follows:

- Critically examine the system in question.
- Divide the system into its various components.
- Examine each individual component and record all of the various ways the components might fail. Rate each potential failure as to the degree of hazards posed (0 = no hazard, 1 = slight, 2 = moderate, 3 = extreme, 4 = severe).
- Examine all potential failures for each individual component of the system and decide what effect the failures could have.

Figure 3.1 is an example of an FMEA conducted on a direct extrusion process. The process/system is broken down into seven components: die backer, die billet, dummy block, pressing stem, container liner and container fillet. The types of failures that might occur are identified as corrosion, cracking, shattering, bending, and surface wear. Of the various components only the dummy block poses an extreme hazard and a corresponding hazard to workers.

As illustrated in figure 3.2 an FMEA produces an extensive analysis of a specific process or system. However, FMEAs have their limitations. First, the element of human error is missing. This is a major weakness because human error is more frequently at the heart of a workplace accident than is system process failure. This weakness can be overcome by coupling human error analysis, which is covered later in this chapter. Second, FMEAs focus on the components of a given system as if the components operate in a vacuum. They do not take into account the interface mechanisms between components or between systems. It is at these interface points that problems often occur.

➥ Hazard and Operability Review (HAZOP) Technique

HAZOP is an analysis method that was developed for use with new processes in the chemical industry. Its strength is that it allows

Plastic Extrusions, Inc.
17, Industrial Boulevard
Ford Walton Beach, Florida 32548

Department Manufacturing Process/System Direct Extrusion Date November, 12 1992

Component	Type of Potential Failure	Potential Effect On: Component	Related Component	Process/ System	Workers	0	1	2	3	4	H	M	L	U	Exam-ination Method	Recomm-endation
Diebacker	Corrosion	Shutdown to replace	None	Shutdown to replace	None	✓							✓		Visual	Periodic checks for corrosion
Die	Cracking	Shutdown to replace	Damage to backer	Shutdown to replace	None			✓				✓			Visual	Periodic checks for cracks
Billet	—	—	—	—	—	–	–	–	–	–	–	–	–	–	—	—
Dummy block	Shaftering	Shutdown to replace	Could damage others	Shutdown all	Injuries from flying				✓			✓			Visual	Inspect and replace periodically
Pressing stem	Bending	Shutdown to replace	None	Shutdown to replace	None		✓						✓		Visual	Inpect and replace periodically
Container liner	Surface wear	Shutdown to replace	None	None	None		✓							✓	Visual	Periodic checks for wear
Container fillet	—	—	—	—	—	–	–	–	–	–	–	–	–	–	—	—

Analysis Conducted by : ________ ________ ________ ________

Fig. 3.1. Sample FMEA

Plastic Extrusions, Inc.
17, Industrial Boulevard
Ford Walton Beach, Florida 32548

Department Composites Process/System Plastic Mix Date January 7, 1993

System/Process Component	Factor Analyzed	Guide-word	Resulting Difference	Potential Problems	Recommended Remedy
Flow-gate number one	Amount of flow	No	No Flow	No mix	Make sure flow-gate is open
		Less	Insufficient flow	Weak mix	Troubleshoot and repair the flow-gate
		More	Excess flow	Too strong mix	Troubleshoot and repair the flow-gate

Fig. 3.2. Sample HAZOP

problems to be identified even before a body of experience has been developed for a given process/system. Although originally intended for use with new processes it need not be limited to new operations. HAZOP works equally well with old process/systems.

HAZOP consists of forming a team of experienced, knowledgeable people from a variety of backgrounds relating to the process/system. The team members brainstorm about potential hazards. The health and safety professional should chair the team and serve as a facilliator. The chairman role is to draw out and record the ideas of team members; make sure that one member does not dominate or intimidate other members; encourage maximum participation from all members; and assist members in combining ideas where appropriate to form better ideas.

A variety of approaches can be used with HAZOP. The one recommended by the American Institute of Chemical Engineers (AICHE) is probably the most widely used. AICHE recommends the following guide words; *no, less, more, part of, as well as, reverse, and other than.*

These guide words relate to the operation of a specific component in the system or a specific part of an overall operation. They describe ways in which the component might deviate from its design or its intended mode of operation. For example, if a component should rotate 38 degrees in a cycle and it fails to rotate at all, the *no* guide word applies. If it rotates fewer than 38 degrees, the *less* guide word applies. *More* would apply if the component's rotation exceeded 38 degree. *Reverse* would be used if the component rotated 38 degrees in a direction opposite of the one intended. As *well* as is similar to *more* in that it indicates an increase in an intended amount. *Other than* is used when what actually occurs is something completely different from what was intended. For example, if the component fell off rather than rotating 38 degrees, *other than* would be used.

A HAZOP proceeds in a step-by-step manner. These steps are summarized as follows:

1. Select the process/system to be analyzed.
2. Form the team of experts.
3. Explain the HAZOP process to all team members.
4. Establish goals and time frames.
5. Conduct brainstorming sessions.
6. Summarize all input.

Figure 3.3 is an example of a form that can be used to help organize and focus brainstorming sessions. It can also be used for summarizing the results of the brainstorming sessions. This particular example involves a plastic mixing process. Only one component in the process (flow gate number 1) has been analyzed. If the flow gate does not work as intended, there will be no flow, too little flow, or too much flow. Each condition will result in a specific problem. Action necessary to correct each situation has been recommended. Every critical point, sometimes referred to as a node, in the process would be analyzed in a similar manner.

HAZOPs have the same weaknesses as FMEAs: They predict problems associated with system/process failures. However, these are technological failures. HAZOPs do not factor human error into the equation. Since human error is so often a factor in accidents, this is a weakness that must be dealt with. The next section sets forth guidelines for analizing human error.

➭ Human Error Analysis (HEA) Technique

HEA, as described in the context of this chapter, is used to predict human error and not as an after-the-fact process. Although the records of past accidents can be studied to identify trends that can, in turn, be used to predict accidents. This should be done as part of an accident investigation. HEA should be used to identify hazards before they cause accidents.

Two approaches to HEA can be effective: (*i*) observing employees at work and noting hazards (the task analysis approach); and (*ii*) actually performing job tasks to get a first-hand feel for hazards. Regardless of how the HEA is conducted, it is a good idea to perform it in conjunction with FMEAs and HAZOPs. This will enhance the effectiveness of FMEAs, HAZOPs, and HEAs.

➭ Technique of Operation Review (TOR)

TOR is an analysis method that allows supervisors and employees to work together to analyze workplace accidents, failures, and incidents. It answers the question, "Why did the system allow this incident to occur?" Like FMEA and HAZOP, this approach seeks to identify systemic causes. It does not seek to assign blame.

TOR is not new. It was originally developed in the early 1970s by D.A. Weaver of the American Society of Safety Engineers. However, for twenty years user documentation on TOR was not readily available. Consequently, widescale use did not occur until the early 1990s when documentation began to be circulated. Richard G.Hallock describes TOR as follows:

TOR is a hands-on analytical methodology designed to determine the root causes of an operation failure. Because it uses a work sheet written in simple-to-understand terms and follows an uncomplicated Yes/No decision-making sequence, it can be used even at the lowest levels of the firm. TOR is triggered by an incident occurring at a specific time and place and involving specific people. It is not a hypothetical process. It demands careful and systematic evaluation of the real circumstances surrounding the incident, and results in isolating the specific ways in which the organization failed to prevent the occurrence.

A weakness of TOR is that it is designed as an after-the-fact process. It is triggered by an accident or incident. A strength is its involvement of line personnel in the analysis. The process proceeds as follows:

1. Establish the TOR team. It should consist of workers who were present when the accident/incident occurred, the supervisor, and the health and safety professional. The health and safety professional should chair the team and serve as a facillitator.
2. Conduct a round table discussion to establish a common knowledge base among team members. At the beginning of the discussion, five team members might have five different versions of the accident/incident. At the end, there should be consensus.
3. Identify one major systematic factor that led to or played a significant role in causing the accident/incident. This one TOR statement, about which there must be consensus, serves as the starting point for further analysis.
4. Use the group consensus to respond to a sequence of yes/no options. Through this process the team identifies a number of factors that contributed to the accident/incident.
5. Evaluate identified factors carefully to make sure there is team consensus about each. Then, prioritize the contributing factors beginning with the most serious.
6. Develop corrective/preventive strategies for each factor. Include them in a final report that is forwarded through normal channels for appropriate action.

Fault Tree Analysis (FTA) Technique

FTA can be used to predict and prevent accidents or as an investigative tool after the fact. FTA is an analytical methodology that uses a graphic model to visually display the analysis process. A fault tree is built of special symbols, some derived from Boolean algebra. Consequently, the resultant model resembles a logic diagram or a flow chart. Figure 3.4 describes the symbols used in constructing fault trees. Figure 3.5 shows how these symbols might be used to construct a fault tree. The top box in a fault tree represents the accident/incident that either could occur or has occurred.

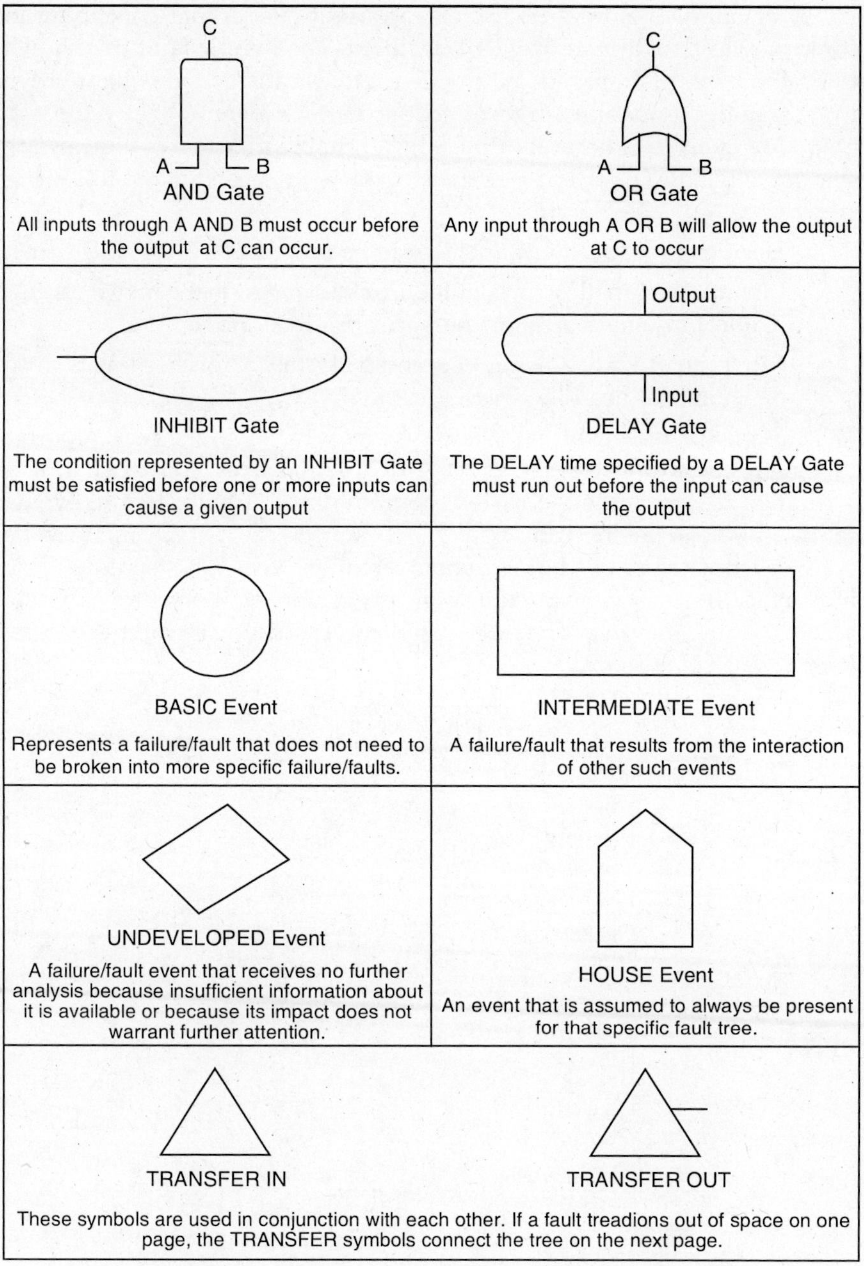

Fig. 3.3. Symbols used in fault tree analysis.

All symbols below the top box represent events that contribute in some way to the ultimate accident/incident. The sample fault tree shown in figure 3.4 is qualitative in nature. Fault trees can be made quantitative by assigning probability figures to the various events below the top box. However, this is rarely done since reliable probability figures are seldom available. A fault tree is developed as explained in the following steps:

1. Decide on the accident/incident to be placed at the top of the tree.
2. Identify the broadest level of failure/fault event that could contribute to the top event. Assign the appropriate symbols.
3. Move downward through successively more specific levels until basic events are identified.

Experience, deliberate care, and systematic analysis are very important in constructing fault trees. Once a fault tree has been constructed, it is examined to determine the various combinations of failure/fault events that could lead to the top event. With simple fault trees this can be accomplished manually, but with more complex trees this step is difficult. However, there are now computer programs available to assist in accomplishing this step. The final step involves making recommendations for preventive measure.

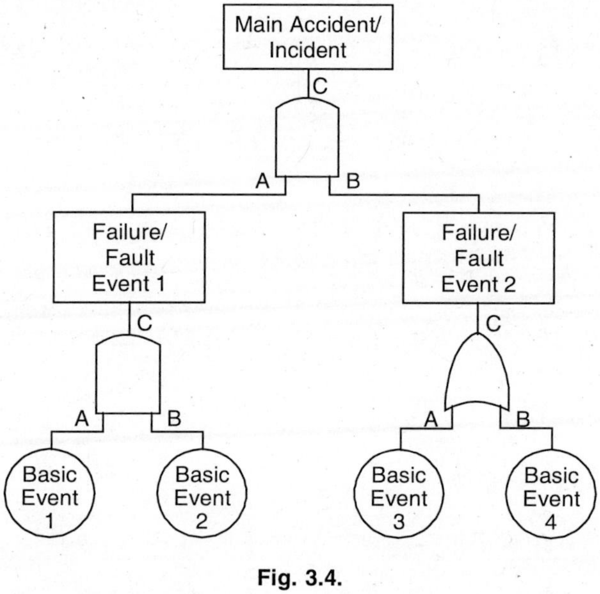

Fig. 3.4.

➷ Risk Analysis Technique

Where are we at risk? where are we at greatest risk? These are important questions for health and safety professionals involved in analyzing the workplace for the purpose of identifying and overcoming hazards. *Risk analysis* in an analytical methodology normally associated with insurance and investments. However, risk analsysis can be used to analyze the workplace, identify hazards, and develop strategies for overcoming these hazards. The risk analysis process focuses on two key questions:

1. How *frequently* does a given event occur?
2. How *severe* are the consequences of a given event?

The fundamental rule of thumb of risk analysis is that *risk is decreased by decreasing the frequency and severity of hazard-related events*.

Health and safety professionals should understand the relationship that exists between the frequency and severity factors relating to accidents. Historical data on accidents, injuries, and illness show that the less severe an injury/illness the more frequently it will occur. Correspondingly, the more severe an injury/illness the less frequently it is likely to occur. For example, there are many more minor scrapes, bumps, and abrasions experienced in the workplace than major debilitating injuries such as amputations or broken bones.

A number of different approaches can be used in conducting a risk analysis. One of the most effective is that developed by Chapanis Chapanis's approach to risk analysis considers both probability and impact.

Probability levels and corresponding frequency of occurrence ratings are as follows: 1 = Impossible (frequency of occurrence > 10^{-8}/day); 2 = Extremely unlike (frequency of occurrence > 10^{-6}/day); 3 = Remote (frequency of occurrence > 10^{-5}/day); 4 = Occasional (frequency of occurrence > 10^{-4}/day); 5 = Reasonably probable (frequency of occurrence > 10^{-3}/day; and 6 = Frequent (frequency of occurrence > 10^{-3}/day).

The lowest rating (1) means is it impossible that given error will be committed or a given failure will occur. The highest rating (6) means it is very likely that a given error will be committed frequently or a given level of probability. For example, the expected frequency of occurrence for a probability level of remote is ten to the negative fifth power per day.

Severity levels can also be rated, with the likely consequence of an accident failure event of that severity. The least severe incidents (1) are not likely to cause an injury or damage property. The most severe incidents (4) are almost certain to cause death or serious property damage. Critical accidents (3) may cause severe injury or major loss. Marginal accidents (2) may cause minor injury, minor occupational illness, or minor damage.

3.3. RECOGNIZING AND CONTROLLING HAZARDS

Industrial hygienists recognize that engineering, work practice, administrative controls are the primary means of reducing employee exposure to occupational hazards.

Engineering Hazard Control: "Engineering hazard controls" may be defined as an installation of equipment, or physical facilities including, if necessary, the selection and arrangement of experimental equipment. Engineering controls remove the hazard, either by initial design specifications or by applying methods of substitution, minimization, isolation, or ventilation. Engineering controls are the most effective hazard control methods, especially when introduced at the conceptual stage of planning when control measure can be integrated more readily into the design. They tend to be more effective than other hazard controls (administrative controls and personal protective equipment) because they remove the source of the hazard or reduce the hazard rather than lessen the damage that may result from the hazard. They are also less dependent on the chemical user who, unfortunately, is subjected to all of the frailties which befall humans (*e.g.*, forgetfulness, preoccupation, insufficient knowledge).

Substitution refers to the replacement of a hazardous material or process with one that is less hazardous (*e.g.*, the replacement of mercury thermometers with alcohol thermometers or dip coating materials rather than spray coating to reduce the inhalation hazard). **Minimization** is the expression used when a hazard is lessened by scaling down the hazardous process. Hence, the quality of hazardous materials used and stored is reduced, lessening the potential hazards (*e.g.*, using micro-scale glassware so that chemical reactions can be carried out on a smaller scale). **Isolation** is the term applied when a barrier is interposed between a material, equipment or process hazard and the property or persons who might be affected by the hazard (*e.g.*, glove box, blast shield).

Substitution is usually the least expensive and the most positive method of controlling hazards and should always be the first engineering hazard control measure considered. Minimization should always be the next engineering control measure attempted after examining substitution followed by the consideration of isolation. Isolation is particularly useful when the material, equipment or process requires minimal contact or manipulations. When these previously mentioned control methods are not feasible, ventilation is the next desirable engineering option.

Ventilation is used to control toxic and/or flammable atmospheres by exhausting or supplying air to either remove hazardous atmosphere at their source or dilute them to a safe level. The two types of ventilation are typically termed **local exhaust** and **general ventilation**. Local exhaust attempts to enclose the material, equipment or process as much as possible and to withdraw air from the physical enclosure at a rate sufficient to assure that the direction of air movement at all openings is always into the enclosure (*e.g.*, fumehood). General ventilation attempts to control hazardous atmospheres by diluting the atmosphere to a safe level by either exhausting or supplying air to the general area (*e.g.*, evaporative cooler).

Local exhaust is always the preferable ventilation method but is typically more costly. For some situation, general ventilation may suffice but only if the following criteria are met: only small quantities of air contaminants are released into the area at fairly uniform rates; there is sufficient distance between the person and the contaminant source to allow sufficient air movement to dilute the contaminant to a safe level; only materials of low toxicity or flammability are being used; there is no need to collect or filter the contaminant before the exhaust air is discharged into the environment (including the rest of the building), and the contaminant will not produce corrosion or other damage to equipment in the area or in any way affect other building occupants outside the general use area (*e.g.*, foul odors).

2. Work Practice Controls

Alter the manner in which a task is performed. Some fundamental and easily implemented work practice controls include (1) changing existing work practices to follow proper procedures that minimize exposures while operating production and control equipment; (2) inspecting and maintaining process and control equipment on a regular basis; (3) implementing good housekeeping procedures; (4) providing good supervision; and (5) mandating that eating, drinking, smoking, chewing tobacco or gum, and applying cosmetics in regulated areas be prohibited.

3. Administrative Control

It consists of managerial efforts to reduce hazards through planning, information and training (e.*g*. hazard communication), written policies and procedures (*i.e.*, the chemical hygiene plan), safe work peactices and environmental and medical susveillance *e.g*. workplace inspections equipment preventive maintenance, and exposure monitoring.

4. Personal Protective Equipment

When effective work practices or engineering controls are not feasible or while such controls are being instituted, appropriate personnel protective equipment are used. Personal protective equipment include articles to protect the eyes, skin and respiratory tract (*e.g.*, goggles, face shield, coats, gloves, aprons, respirators).

✎ IMPORTANT QUESTIONS

1. Define industrial hazard. What are its type.
2. Discuss the various hazard identification techniques.
3. Explain 'FMEA' and 'HAZOP' techniques of hazard identification.
4. How hazards can be prevented.
5. Explain accident investigation, reporting and analysis.

❑❑❑

CHAPTER 4

PLANNING FOR SAFETY

Alice asked at the crossroad, "which road should I take?" "Where do you want to go?" asked the Mad Hatter. "I don't know," said Alice. "Then it doesn't matter which road you take" he replied. Alice in wonderland.

4.1. PURPOSE OF PLANNING

Planning is an essential forerunner of action. It is a process of deciding in advance the future course of action. A prerequisite for achieving industrial safety is need of planning. Planning forms the backbone of any management activity including industrial safety management. Although often managers avoid this, preferring the excitement of activity to the inactivity of thought.

At this juncture, let us consider the example of a well-known industrial conglomerate which decided to construct simultaneously three identical chemical plants in India, the United Kingdom and Japan. Indians completed the plant, Britishers were at the verge of completion but Japanese were yet to lay the foundation. Indian plant started production, British plant was about to start production but Japanese were far behind. However, before the Indian plant came into full production, problems began to emerge. These were dismissed as mere "teething troubles". In due course, such problems started appearing in UK plant when it started production. Meanwhile, after inordinate planning, Japanese construction started and the project therefore progressed flawlessly. The Indian plant continued giving problems and finally stopped production needing a major overhaul. The British plant also by then started giving major problems. Both Indian and British plants had to undergo major replanning. Through careful planning, the Japanese plant got completed on schedule and as per plan. Through careful planning, the Japanese plant became fully operational without any problem. This example shows the relevance of systematic planning in industries and business. Japanese go by systematic planning slowly but steadily when compared to the rest of the world and achieve brilliant success.

Planning involves decision making in advance taking due account of the constraints and priorities, resource available and shape of things to come in future. Anticipation of goals, choosing between alternatives, the possible pitfalls and lacuna during working, any bottlenecks in the smooth functioning of various activities are some of the main concerns of planning. Clarity of vision, intellectual acumen and a strong but conscious determination pave the way to success so far as planning is concerned.

4.2. PLANNING PROCEDURE

To start with the planning process for safety, the management should decide the organisational plans for safety. Everyone in the organisation should understand the importance of safety practices and how to incorporate them in daily work schedule. Some safety standards should be agreed upon and understood by all. Everyone with the organisation should strive continuously to satisfy those standards. The senior management of the organisation should act in unison to give a clear direction for the achievement of safety and concentrate its efforts on the key measures of safety.

Some basic considerations while planning for safety are as follows:

❑ Involvement of Workers

Workers should be motivated to be safety conscious. Right type of training must be imparted to workers after considering their socio–economic backgrounds, and their attitudes towards safety.

❑ Age and other Traits of Workers

The age of the workers as well as their salient traits have some bearing on safety. Since the age as well as traits of the workers must have due consideration for building up strategy for safety management.

❑ Experience of Workers

The experience of workers in the field plays a pivotal role in preventing hazards to a great extent. The novice among the workers should have proper interaction with the experienced worker so that adequate knowledge may be disseminated among the new entrants.

❑ Actual Working Procedures

A safe layout and working procedure contribute individually or collectively in the field of safety management.

❐ Working Conditions

Poor working conditions, worn out machines, lack of repair and maintenance and an uncongenial working environment have deleterious effects in the field of safety management.

❐ Past experience of Planning

Past experience of planning plays a very important role in chalking out future line of action in safety planning. The successful as well as the unsuccessful experiments of the past can help to formulate superior planning standard.

❐ Role of Employer or Management

The role of employer or management in inculcating the norms of safety among workers is a very important consideration for safety planning. Mutual cooperation and understanding between the management and the workers is a touchstone for a successful planning.

- The outcomes from such a planning process include a basic understanding of purpose, nature and scope of safety in an organization.
- an agreed company safety mission statement.
- allocation of roles and responsibilities to each member of safety management team.
- implementation of safety plan together with an agreed communication.
- dates for series of review meetings.

4.3. ESSENTIALS OF PLANNING PROCESS

A formal safety planning provides many benefits including: systematic thinking, better coordination, sharper objectives, improved performance standards and management involvement. All these result in a planned approach to tackle safety related problems that can help in higher productivity. If we don't have a good planning process, our employees will design their own and that may not be the best. A good planning process needs to exist throughout an organisation and should be a standard process. Everybody should be trained for it. Once in place, it provides a common planning language with common formats.

4.4. RANGE OF PLANS

Safety plans range from short to medium to long. A long range safety plan includes an analysis of the current situation and then defines the broad objectives and strategies that must be. Long range plans may be of varied time durations, generally 3 – 5 years. The medium range of safety plans includes tactical decisions on safety rules and regulations. It ranges from 1 – 2 years. Short range plans deal with operational or day-to-day decisions regarding safety related problems. These plans range is 1 year or less.

4.5. SAFETY POLICY

A safety policy provides the foundation for every successful safety programme and could help to avoid the expenses, inconvenience and other consequences of workplace accidents by making sure that employees know what is expected of them. A written safety policy is the beginning of any safety programme. A policy is a basic guideline which dictates the thinking style as well as the actions to achieve the desired goals or objectives. policies also restrict the arena within which decisions are to be taken in conformity with the objectives. A safety policy provides a suitable base for coordination of safety activities in the various levels of hierarchies. The various elements of a safety policy are:

- objectives of a safety programme
- persons responsible for various safety practices.
- consequences of not following established procedure.
- reporting of unsafe conditions.
- reporting of accidents.
- list of specific safety rules and regulations.
- list of safety hazard to look for.
- safety clothing and equipment.
- use of seat belts.
- good housekeeping standards.

Therefore principles, rules and norms either explicit or implicit set up by the management and others along with the constraints therein fall under the ambit of policies.

4.6. FORMULATION OF SAFETY POLICY

The formulation of safety policy includes ways and means to make safety policy positive or effective. An effective safety policy need to be flexible. It should be flexible and broad-based in the sense that they will provide guidelines for problems which vary in nature such that frequent changes in the policy are avoided. The safety policy should be so formulated that explicitness is maintained. This implies that the policy should be in writing and ambiguous terms or statements should be absent in the formulation of the said policy. It should be coherent neither contradictory not at variance with other policies thus maintaining the very essence of compatability. The safety policies must reflect pragmatism. The viability and the success of the policy lie in the input-output assessment such that the benefits are commensurate with monetary involvement. Any safety policy must have clear acceptability among the workers and should have enough credibility and sincerity of purpose so that workers do not show a sense of aversion towards the policy.

While formulating a safety policy the following points should be kept in mind:

- The target in the field of safety must be clearly defined by the management.
- The powers vested with the authority must be defined in unambiguous terms.
- The norms and standards should be clearly defined.
- The duties and responsibilities of safety personnel must be well defined so that no confusion arises on this account. Accountability is the prime need in all safety management activities. Scope of various personnel in respect of safety must be well defined in order to have a foolproof safety management system.

Health and safety at work Act 1974, contained various ways and means to make safety policy effective. Some of the guidelines for the formulation of an effective safety policy are:

- Formulation of self regular safety policy by the employees should be allowed.Workers can be informed orally of safety standards and procedures, but for lasting impact there is no substitute for a written policy to which an employee may refer. The policy statement sets the tone and indicates what the consequences will be. A written policy reflect the purpose of safety program and will be assigned safety responsibilities and what their responsibilities and accountability required will be
- There are specific statements and elements the safety policy should contain. They will help to not only reflect your philosophy and the intent of safety policy, but also give specific information and guidance to employees about safety.

4.7. ELEMENTS OF SAFETY POLICY

A safety policy should contain:

- a strongly worded affirmative statement that acknowledges and support the company's responsibility.

- titles; positions and locations of those responsible for policy design, enforcement and modification.
- a system for any employee to report violations of safety rules confidentially.
- documentation that establishes that the policy was given to all employees.
- a statement that supervisors are held to act as agents for the company.
- a statement that any deviation from known policy by supervisors will not be tolerated.
- a process to address how questions are to be answered and emergencies handled.
- the penalties for employees who don't follow safety rules.
- employees should be convinced regarding the need and usefulness of the safety policy. This goal can be achieved by ensuring that the workplace as well as the working environment are free from risks.
- workers should be taken into confidence and their involvement in the framing of safety policy and decision-making should be welcomed.
- the safety policy shall aim at rewarding persons who exhibit exemplary performance in containing accidents.
- safety policy shall be positive and optimistic in approach. The usefulness of safety norms should be highlighted through various information, data and other statements.
- safety policy should be specific though flexible in case of certain circumstances.
- while ravening the safety policy the workers should be taken into confidence.
- no relaxation should be allowed in the enfocemat of safety policy.

4.8. IMPLEMENTATION OF SAFETY POLICY

The implementation phase involves development of an implementation plan. The implementation plan lists detailed steps or tactics necessary to accomplish the safety objective. Hence, implementation plans are typically prepared at the lower management levels. Special care is needed in implementing the plan which is most crucial and difficult portion of the planning process.

✎ IMPORTANT QUESTIONS

1. Define safety planning.
2. Discuss the purpose of planning industrial safety.
3. Explain the procure of safety planning.
4. What are the considerations while planning for safety?
5. What can be the benefits of planning industrial safety?
6. List the range of safety plan.
7. What is a safety policy.
8. Explain the process of policy formulation.
9. What are the elements of safety policy?
10. How a safety policy can be implemented.

❑❑❑

CHAPTER 5

Safety Standards and Legislation

5.1. SAFETY STANDARDS

B65 Standards

- **B65.1 - 1995 Safety standard-printing press systems.** This standard provides safety specifications for the design of the press drive controls, safety signaling system and guarding for web and sheet-fed printing press. Also includes additional user safety recommendations.
- **B65.2 - 1999 Safety standard - Binding and finishing system.** This standard provides safety specifications for the design of binding and finishing system drive controls, safety signaling systems, mechanical safety devices, and includes safe operating guidelines.
- **B65.3 - 2001 Safety standard - Guillotine paper cutters, mill trimmers and integral handling equipment** This standard specifies operational & mechanical safety specifications for the design and use of guillotine cutters, mill trimmers and integral handling equipment, when they are used as intended, under conditions foreseen by the manufactures.
- **B65.4 - 0220 Safety standard - Three-knife trimmers, including rotary, and signle- and multiple-knife trimmers.** This standard specifies operational and mechanical safety specifications for the design and use of stand-alone three-knife trimmers, when they are used as intended and under the conditions foreseen by the manufacturers. It addresses significant mechanical hazards but does not address other hazards such as shock, explosion, fire, noise/sound levels or exposure to chemicals. This standard does not address all hazards that may exist during maintenance operations. For maintenance operations, OSHA lockout/tagout regulations may apply.

- **B65.5 - 1996 Safety standard-Stand alone platen presses.**This standard provides operational and mechanical safety specifications for the design and use of webfed and sheetfed stand-alone platen press systems intended for diecutting, embossing, foil stamping and/or printing of paper, board and other materials processed in a similar manner.

ISO Standards

- **ISO 9355-1:1999 Ergonomic requirements for the design of displays and control actuators - Part 1: Human interactions with displays and control actuators.** This International Standard applies to the design of displays and control actuators on machinery. It specifies general principles for human interaction with displays and control actuators, to minimize operator errors and to ensure an efficient interaction between the operator and the equipment. It is particularly important to observe these principles when an operator error may lead to injury or damage to health.

- **ISO 9355-2:1999 Ergonomic requirements for the design of displays and control actuators - Part 2: Displays.** This International Standard gives guidance on the selection, design and location of displays to avoid potential ergonomic hazards associated with their use. It specifies ergonomics requirements and covers visual, audible and tactile displays, and applies to displays used in machinery (e.g. devices and installations, control panels, operating and monitoring consoles) for occupational and private use.

- **ISO 11553:1996 Safety of machinery - Laser processing machines - Safety requirements.** This International Standard describes hazards generated by laser processing machines and specifies safety requirements relating to radiation hazards and hazards generated by materials and substances. It also specifies information to be supplied by manufacturers of such equipment. Not applicable to laser products or equipment manufactured solely for photo lithography, stereolithography, holography, medical applications or data storage.

- **ISO/TR 11688-1:1995 Acoustics-Recommended practices for the design of low-noise machinery and equipment - Part 1: Planning.** This International Technical Report is an aid to understanding the basic concepts of noise control in machinery and equipment. The practice presented is intended to assist designers at any design stage to control the noise of the final product. Reference is made to numerous technical publications dealing with acoustical problems.

- **ISO/TR 11688-2:1998 Acoustics - Recommended practice for the design of low-noise machinery and equipment - Part 2.** Introduction to the physics of low-noise design. This Technical Report provides the physical background for the low-noise design rules. It is intended for use by designers of machinery and equipment as well as users and/or buyers of machines and authorities in the field of legislation, supervision or inspection. Equations given herein will improve the general understanding of noise control. In many cases they allow a comparison of different versions of design, but are not useful for the prediction of absolute noise emission values.
- **ISO 12100-1:2003 Safety of machinery - Basic concepts, general principles for design - Part 1: Basic terminology, methodology.** This International Standard defines basic terminology and methodology used in achieving safety of machinery. The provision stated herein are intended for the designer. It does not deal with damage to domestic animals, property or the environment.
- **ISO 12100-2:2003 Safety of machinery - Basic concepts, general principles for design - Part 2: Technical principles.** This International Standard defines technical principles to help designers in achieving safety in the design of machinery. It is intended to be used together with ISO 12100-1 when considering the solution to a specific problem. The two parts of ISO 12100 can be used independently of other documents or as a basis for the preparation of other type-A standards or type-B or C standards. This standard does not deal with damage to domestic animals, property or the environment.
- **ISO 12648:2003 Graphic technology - Safety requirements for printing press systems.** This International Standard applies to printing press system, including auxiliary equipment and finishing machines, in which all the machine actuators (e.g. drives) of the equipment in the system are controlled by the same control system. It applies only to systems in which a printing press is part of the system. It addresses recognized hazards specific to printing press systems in the following areas: mechanical; electrical; slipping, tripping, falling; ergonomics; noise; radiation; fire and explosion; thermal; and emissions. It applies to new machines manufactured after December 31 of the year following the year of publication.
- **ISO/TS 13732-2:2001 Ergonomics of the thermal environment - Methods for the assessment of human responses to contact with surfaces - Part 2: Human contact with surface at moderate temperature.**

This part of ISO/TS 13732 presents principles and methods for predicting thermal sensation and degree of discomfort in cases where parts of the body contact solid surfaces at moderate temperatures. Also deals with thermal sensation for contacts of hands, feet and for sitting position on the floor.

- **ISO13849-1:1999 Safety of machinery - Safety-related parts of control systems - Part 2: Validation.** This International Standard specifies the procedures and conditions to be followed for the validation by analysis and testing of the safety functions provided and the category achieved for the safety-related parts of the control system in compliance with EN 954-1 (ISO 13849-1), using the design rationale provided by the designer. This International Standard does not give complete validation requirements for programmable electronic systems and therefore can require the use of other standards.
- **ISO/TR 13849-100:2000 Safety of machinery - Safety-related parts of control systems - Part 100: Guidelines for the use and application of ISO 13849-1.** This Technical Report provides guidance on the appropriate use and interpretation of ISO 13849-1:1999. It also gives further information on how the control system contributes to reducing risk in the machine; what is meant by the safety-related parts of the control system in relation to safety functions; the proper selection and use of categories; and the role of annex. B of ISO 13849-1:1999.
- **ISO 13850:1996 Safety of machinery - Emergency stop - Principles for design.** This International Standard specifies functional requirements and design principles for the emergency stop of machinery, independent of the type of energy used to control the function. This International Standard does not deal with functions such as reversal or limitation of motion, deflection, shielding, braking, or disconnecting, which may be part of the emergency stop function.
- **ISO 13851:2002 Safety of machinery - Two-hand control device - Functional aspects and design principles.** This International Standard specifies the safety requirements of a two-hand control device and the dependency of the output signal from the input signals. It describes the main characteristics of two-hand control devices for the achievement of safety and sets out combinations of functional characteristics for three types. It provides requirements and guidance on the design and selection of two-hand control devices including their assessment, the prevention of defeat and the avoidance of faults. It also provides requirements and guidance for two-hand control devices containing a programmable electronic system.

- **ISO 13852:1996 Safety of machinery - Safety distances to prevent danger zones being reached by the upper limbs.** This International Standard establishes values for safety distances to prevent danger zones being reached by the upper limbs of persons of 3 years of age and above. Distances apply when adequate safety can be achieved by distances alone.
- **ISO 13853:1998 Safety of machinery - Safety distances to prevent danger zones being reached by the lower limbs.** This International Standard establishes values for safety distances to prevent access and distances to impede free access to machinery danger zones to prevent their being reached by the lower limbs of persons 14 years of age and above.
- **ISO 13854:1996 Safety of machinery - Minimum gaps to avoid crushing of parts of the human body.** The object of this International Standard is to enable the user (e.g. standard makers, designers of machinery) to avoid hazards form crushing zones. It specifies minimum gaps relative to parts of the human body and is applicable when adequate safety can be achieved by this method.
- **ISO 13855:2002 Sefety of machinery - Positioning of protective equipment with respect to the approach speeds of parts of the human body.** This International Standard provides parameters based on values for hand/arm and approach speeds and the methodology to determine the minimum distances form sensing or actuating devices of protective equipment to a danger zone. It does not apply to protective equipment which is intended to be moved, without tools, nearer to the danger zone than the calculated distance, e.g. pendant two-hand control devices.
- **ISO 13856-1:2001 Safety of machinery - Pressure-sensitive protective devices - Part 1: General principles for design and testing of pressure-sensitive mats and pressure-sensitive floors.** This international standard specifies requirements for pressure-sensitive mats and floors normally actuated by the feet, for use as safety devices to protect persons from dangerous machinery. The minimum safety requirements for the performance, marking and documentation are given. It deals with pressure-sensitive mats and floors, regardless of type of energy used, designed to detect persons weighing more than 35 kg and persons weighing more than 20 kg.
- **ISO 14118:2000 Safety of machinery - Prevention of unexpected start-up.** This International Standard specifies built-in safety measures

aimed at preventing unexpected machine start-up to allow safe human interventions in hazard zones.

- **ISO 14119:1998 Safety of machinery - Interlocking devices associated with guards - Principles for design and selection.** This International Standard specifies principles for the design and selection, independent of the nature of the energy source, of interlocking devices associated with guards.
- **ISO 14120:2002 Safety of machinery - Guards - General requirements for the design and construction of fixed and movable guards.** This International Standard specifies general requirements for the design and construction of guards provided primarily to protect persons form mechanical hazards. It applies primarily to machines which will be manufactured after it is published. The requirements are applicable if fixed and movable guards are used, but does not cover guards which actuate interlocking devices. It does not provide requirement for special systems relating specifically to mobility or to the ability to lift loads.
- **ISO 14121:1999 Safety of machinery - Principles of risk assessment.** This International Standard establishes general principles for the procedure known as risk assessment, by which the knowledge and experience of the design, use, incidents, accidents and harm related to machinery is brought together in order to assess the risks during all phases of the life of the machinery. This International Standard gives guidance on the information required to allow risk assessment to be carried out. Procedures are described for identifying hazards and estimating and evaluating risk. This International Standard is not intended to provide a detailed account of methods for analyzing hazards and estimating risk, as this is dealt with elsewhere (e.g. text books and other reference documents).
- **ISO 14122-1:2001 Safety of machinery - Permanent means of access to machinery - Part 1: Choice of fixed means of access between two levels.** This International Standard applies to all machinery (stationary and mobile) where fixed means of access are necessary. It advises on the correct choice of access means when the necessary access to the machine is not possible directly from the ground level or from a floor. Applies to access means which are a part of a machine; means of access specific to the machine which are not permanently fixed to the machine; and may apply to means of access which are part of the building. Not applicable to devices specifically designed to lift persons between two levels.

- **ISO 14122-2:2001 Safety of machinery - Permanent means of access to machinery - Part 2: Working platforms and walkways.** This International Standard applies to all machinery (stationary and mobile) where fixed means of access are necessary. It applies to working platforms and walkways which are a part of a machine; are specific to the machine which are not permanently fixed to the machine; and may apply to working platforms and walkways to part of the building where the machine is installed. Not applicable to devices specifically designed to lift persons between two levels.
- **ISO 14122-3:2001 Safety of machinery - Permanent means of access to machinery - part 3: Stairs, stepladders and guard-rails.** This International Standard applies to all machinery (stationary and mobile) where fixed means of access are necessary. Applies to stairs, step ladders and guard-rails which are part of machine; are specific to the machine which are not permanently fixed to the machine; and may apply to stairs, stepladders and guard-rails to part of the building where the machine is installed.
- **ISO 14123-1:1998 Safety of machinery - Reduction of risks to health form hazardous substances emitted by machinery - Part 1: Principles and specifications for machinery manufacturers.** This part of ISO 14123 deals with principles for the control of risks to health due to hazardous substances from machinery. This part of ISO 14123 is not applicable to substances which are a hazard to health solely because of their explosive, flammable or radioactive properties or their behaviour at extremes of temperature or pressure.
- **ISO 14123-2:1998 Safety of machinery - Reduction of risks to health form hazardous substances emitted by machinery - Part 2: Methodology leading to verification procedures.** This of ISO 14123 defines a procedure which leads to the selection of critical factors relating to emissions of hazardous substances for the purpose of specifying suitable verification procedures. This part of ISO 14123 is intended to be used in conjunction with ISO 14123-1 and relates specifically to clause 8 of that standard.
- **ISO 15534-1: 2000 Ergonomic design for the safety of machinery - Part 1: Principles for determining the dimensionsrequired for openings for whole-body access into machinery.** This part of ISO 15534 specifies the dimensions of openings for whole-body access into machinery as defined in ISO/TR 12100-1. It provides the dimensions to which the values given in ISO 15534-3 are applicable.

It has been prepared primarily for nonmobile machinery; there may be additional specific requirements for mobile machinery. This part of ISO 15534 shows how to combine the anthropometric data with suitable allowances to take these factors into account. Situations where people are to be prevented from reaching a hazard are dealt with in ISO 13852.

- **ISO 15534-2:2000 Ergonomic design for the safety of machinery - Part 2: Principles for determining the dimensions required for access openings.** This part of ISO 15534 specifies the dimensions of openings for access into machinery as defined in ISO/TR 12100-1. It provides the dimensions to which the values given in ISO 15534-3 are applicable. It has been prepared primarily for non-mobile machinery; there may be additional specific requirements for mobile machinery. This part of ISO 15534 shows how to combine the anthropometric data with suitable allowances to take these factors into account. Situations where people are to be prevented from reaching a hazard are dealt with in ISO 13852.

- **ISO 15534-3:2000 Ergonomic design for the safety of machinery - Part 3: Anthropometric data.** This part of ISO 15534 specifies current requirements for human body measurements (anthropometric data) that are required by ISO 15534-1 and ISO 15534-2 for the calculation of access-opening dimensions as applied to machinery. The data are based on information from anthropometric surveys representative of population groups within Europe comprising at least three million people; both men and women. Measurements meet the requirements of ISO 15534-1 and ISO 15534-2.

IEC STANDARDS

- **IEC 50(161):1990, Ed. 1.0 International Eelectrotechnical Vocabulary - Chapter 161: Electromagnetic compatibility.** This standard contains terms and definitions relating to electro-magnetic compatibility.

- **IEC 60079-1:2003, Ed. 5.0 Electrical apparatus for explosive gas atmospheres - Part 1: Flameproof enclosures "d".** This part of IEC 60079 contains specific requirements for the construction and testing of electrical apparatus with the type of protection flameproof enclosure "d", intended for use in explosive gas atmosphere.

- **IEC 60079-2:2001, Ed 4.0 Electrical apparatus for explosive gas atmosphere - Part 2: Pressurized enclosures "p".** This standard contains the specific requirements (supplementary to IEC 60079-0) for construction and testing of electrical apparatus with pressurized enclosures, of protection type "p", intended for use in explosive gas atmospheres. It specifies needs for pressurized enclosures containing a limited release of a flammable substance. Not applicable where containment system may release air with an oxygen content greater than normal, or oxygen with inert gas in proportion greater than 21%, or for pressurized rooms or analyser houses.

- **IEC 600079-5:1997, Ed. 2.0 Electrical apparatus for explosive gas atmospheres - Part 5: Powder filling "q".** This standard contains the specific requirements for construction, testing and marking of electrical apparatus, in whole and in part, and Excomponents with current rated less than or equal to 16 A, and power consumption rated less than or equal to 1000 VA, in the type of protection powder filling "q", intended for use in potentially explosive gas, vapour and mist atmospheres. Supplement to IEC 600079-0.
- **IEC 60079-5:1997, Am. 1 (2003).** Amendment 1 to Electrical apparatus for explosive gas atmospheres - Part 5: Powder filling "q".
- **IEC 60079-6:1995, Ed. 2.0 Electrical apparatus for explosive gas atmospheres - Part 6: Oil immersion "o".** This standard specifies requirements for the construction and testing of oil immersed electrical apparatus, part of apparatus and Excomponents in the type of protection "0", intended for use in potentially explosive gas atmospheres. It is applicable to electrical apparatus and parts of electrical apparatus which are not ignition capable in normal operation. It is a supplement to IEC 79-0. English/French.
- **IEC 60079-7:2001, Ed. 3.0 Electrical apparatus for explosive gas atmospheres - Part 7: Increased safety "e".** This standard gives requirements for the design, construction, testing and marking of electrical apparatus, with a rated value of supply voltage not exceeding 11 kV r.m.s. a.c. or d.c., with type of protection 'e' that does not produce sparks, arcs, or dangerous temperatures in normal operation.
- **IEC 60079-10:2002, Ed. 4.0 Electrical apparatus for explosive gas atmospheres - Part 10: Classification of hazardous areas.**This standard specifies the classification of hazardous areas where flammable gas or vapour risks may arise, in order to permit the proper selection and installation of apparatus for use in such hazardous areas.
- **IEC 60079-11:1999, Ed. 4.0 Electrical apparatus for explosive gas atmospheres - Part 11: Intrinsic safety "i".** This standard specifies the construction and testing of intrinsically safe apparatus, intended

for use in potentially explosive atmospheres and for associated apparatus, intended for connection to intrinsically safe circuits which enter such atmospheres. Contains details of the test apparatus previously published as IEC 60079-3. It is a supplement to IEC 60079-0:1998.

- **IEC 60079-18:1992, Ed. 1.0 Electrical apparatus for explosive gas atmospheres - Part 18: Encapsulation "m".** This standard prescribes the specific requirements for construction and testing of electrical apparatus and Excomponents which have rated voltages not exceeding 11 kV with the type of protection encapsulation "m". Supplement to IEC 79-0 with some exceptions.
- **IEC 60204-1:2000, Ed. 4.1 Safety of machinery-Electrical equipment of machines - Part 1: General requirements.** This standard applies to the application of electrical and electronic equipment and systems to industrial machines not portable by hand while working, including a group of machines working together in a coordinated manner, but excluding higher level systems aspects (e.g., communications between systems).
- **IEC 60417-1:1998 Graphical symbols for use on equipment - Part 1: Overview and application** This standard contains graphical symbols with the corresponding descriptions in numerical order. It also contains an alphabetical index and a graphic survey of the symbols. Additional information is provided to the user by the classifications according to form and function, and the examples of application. Along with IEC 60417-2, this updates and replaces previous versions of IEC 417 and its annexes.
- **IEC 60417-2:1998 Graphic symbols for use on equipment - Part 2: Symbol originals.** This standard contains a presentation of the full size symbol originals for reproduction purposes in numerical order. The position and size of the graphical symbols within the basic pattern can be determined with the help of the transparent sheet of the basic pattern. With IEC 60417-1, this updates and replaces previous versions of IEC 417 and its annexes.
- **IEC 60529:2001, Ed. 2.1 Degrees of protection provided by enclosures (IP Code).** This standard applies to the classification of degrees of protection provides by enclosures for electrical equipment with a rated voltage not exceeding 72,.5 kV.
- **IEC 60825-1:2001, Ed. 1.2 Safety of laser products - Part 1: Equipment classification, requirements and user's guide.** This standard applies to safety of laser products. It is divided into three separate sections:

Section One (General) and the annexes; Section Two (Manufacturing requirements); & Section Three (User's guide).

- **IEC 60947-1:2001, Ed. 3.2 Low-voltage switchgear and controlgear - Part 1: General rules.** This standard applies, when required by the relevant product standard, to switchgear and controlgear and is intended to be connected to circuits, the rated voltage of which does not exceed 1000 V a.c. or 1500 V d.c. This version includes Edition 3 of the IEC 60947-1 issued in 1999, and the amendments issued in 2000 and 2001.

- **IEC 60947-2:2003, Ed.3.0 Low-voltage switchgear and controlgear - Part 2: Circuit-breakers.** This standard applies to circuit-breakers, the main contacts of which are intended to be connected to circuits, the rated voltage of which does not exceed 1000V a.c. or 1500 V d.c.; it also contains additional requirements for integrally fused circuit-breakers.

- **IEC 60947-3:2001, Ed. 2.1 Low-voltage switchgear and controlgear - Part 3: Switches, disconnectors, switch disconnectors and fuse-combination units.** This standard applies to switches, disconnectors, switch-disconnectors and fuse-combination units to be used in distribution circuits and motor circuits of which the rated voltage does not exceed 1000 V a.c. or 1500 V d.c.

- **IEC 60947-4-1:2002, Ed. 2.1 Low-voltage switchgear and controlgear - Part 4-1: Contactors and motor-starters - Electromechanical contactors and motor-starters.** This standard states the characteristics of contactors and starters and associated equipment, the conditions with which contactors or starters shall comply, the tests intended for confirming that these conditions have been met, and the information to be given with the equipment or in the manufacturer's literature. This publication supersedes IEC 60158-1 (1970) and its Amendment 1 (1983), 60158-1C (1982), 60292-1 (1969) and its Amendment 2 (1983), 60292-2 (1970), 60292-3 (1973) and 60292-4 (1975). It should be read in conjunction with IEC 60947-1.

- **IEC 60947-4-2:2002, Ed. 2.1 Low-voltage switchgear and controlgear-Part4-2: Contactors and motor-starters - AC semiconductor motor controllers and starters.** This standard applies to controllers and starters, which may include a series mechanical switching device, intended to be conected to circuits, the rated voltage of which does not exceed 1000 V a.c.

- **IEC 60947-5-1:2--3, Ed. 3.0 Low-voltage switchgear and controlgear-Part5-1: Control circuit devices and switching element-Electro-mechanical control circuit devices.** This standard applies to control circuit devices and switching elements intended for controlling, signaling, interlocking, etc., of switchgear and controlgear.
- **IEC 60947-5-2:2004, Ed. 2.2 (Edition 2:1997 + Am. 1:1999 & Am. 2:2003) Low-voltage switchgear and controlgear - Part 5-2: control circuit devices and switching elements - Proximity switches (includes Amendment 1 & 2).** This standard applies to inductive and capacitive proximity switches that sense the presence of metallic and/or non-metallic objects, ultrasonic proximity switches that sense the presence of sound reflecting objects and photoelectric proximity switches that sense the presence of object.
- **IEC 60947-5-4:2002, Ed. 2.0 Low-voltage switchgear and controlgear - Part 5-4: Control circuit devices and switching elements - methods of assessing the performance of low-energy contacts - Special tests.** This part of IEC 60947 applies to separable contacts used in the utilization area considered such as switching element for control circuits.
- **IEC 60947-5-5:1997, Ed. 1.0 Low-voltage switchgear and controlgear - Part 5-5: Control circuit devices and switching elements - Electrical emergency stop device with mechanical latching function.** This part of IEC 60947-5 provides detailed specifications relating to the electrical and mechanical construction of emergency stop devices with mechanical latching function and to their testing.
- **IEC 60947-6-1:1998, Ed. 1.2 Low-voltage switchgear and controlgear - Part 6-1: Multiple function equipment-Automatic transfer switching equipment.** This standard applies to Automatic Transfer Switching Equipment (ATSE) to be used in emergency power systems with interruption of the supply to the load during transfer, the rated voltage of which does not exceed 1,000 V a.c. or 1,500 V d.c. It covers ATSE provided with or without an enclosure.
- **IEC 60947-6-2:2002, Ed. 2.0 Low-voltage switchgear and controlgear - Part 6-2: Multiple function equipment - Control and protective switching devices (or equipment) (CPS).** This section of IEC 60947-6 applies to control and protective switching devices (or equipment) (CPS), the main contacts of which are intended to be connected to circuits of rated voltage not exceeding 1000 V a.c. or 1500 V d.c.
- **IEC 60947-7-1:2002, Ed. 2.0 Low-voltage switchgear and controlgear - Part 7-1: Ancillary equipment - Terminal blocks for copper**

conductors. This standard specifies requirements for terminal blocks with screw-type or screwless type terminals primarily intended for industrial or similar use and to be fixed to a support to provide electrical and mechanical connection between copper conductors.

- **IEC 60947-7-2:2002, Ed. 2.0 Low-voltage switchgear and controlgear - Part 7-2: Ancillary equipment - Protective conductor terminal blocks for copper conductors.** This section of IEC 947-7 applies to protective conductor terminal blocks with PE function up to 120 mm^2 (250 MCM) and to protective conductor terminal blocks with PEN function equal to and above 10 mm^2 (AWG 8) with screw-type or screwless-type clamping units, primarily intended for industrial applications.
- **IEC 61010-1:2001, Ed. 2.0 Safety requirements for electrical equipment for measurement, control, and laboratory use - Part 1: General requirements.** This standard specifies general safety requirements for electrical equipment intended for professional, industrial process, and educational use, which may incorporate computing devices: electrical test and measurement equipment; electrical control equipment; electrical laboratory equipment; or accessories intended for use with them, used under specified environmental conditions.
- **IEC 61310-1:1995, Ed. 1.0 Safety of Machinery - Indication, Marking and Actuation - Part 1: Requirements for visual, auditory, and tactile signals.** This part of IEC 1310 specifies requirements for visual, auditory and tactile methods of indicating safety-related information, at the man-machine interface and to exposed persons.
- **IEC 61310-2:1995, Ed. 1.0 Safety of Machinery - Indication, Marking and Actuation - Part 2: Requirements for marking.** This part of IEC 1310 specifies requirements for the marking of machinery. It gives general rules on marking for identification of machinery, for safe use related to mechanical and electrical hazards, and for the avoidance of hazards arising from incorrect connections.
- **IEC 61310-3:1995, Ed. 1.0 Safety of Machinery - Indication, Marking and Actuation - Part 2: Requirements for the location and operation of actuators.** This part of IEC 61310 specifies safety-related requirements for actuators, operated by the hand or by other parts of the human body, at the man-machine interface. It gives general requirements for the standard direction of movement for actuators; the arrangement of an actuator in relation to other actuators; the correlation between an action and its final effects. It is based on IEC 60447, but is also applicable to non-electrotechnical technologies, such as mechanical

and fluid-powered systems. It covers single actuators as well as groups of actuators forming part of an assembly.

- **IEC 61496-1:2004, Ed. 2.0 Safety of machinery - Electro-sensitive protective equipment - Part 1: General requirements and tests.** This part of IEC 61496 specifies general requirements for the design, construction and testing of electro-sensitive protective equipment (ESPE) designed specifically to detect persons as part of a safety related system.
- **IEC 61496-2:1997, Ed. 1.0 Safety of machinery - Electro-sensitive protective equipment - part 2: Particular requirements for equipment using active opto-electronic protective devices (AOPDs).** This part of IEC 61496 specifies requirements for the design, construction and testing of electro-sensitive protective equipment (ESPE) for the safeguarding of machinery, employing active opto-electronic protective devices (AOPDs) for the sensing function.
- **IEC 61508-3:1998, Ed. 1.0 Functional safety of electrical/ electronic/ programmable electronic safety-related systems - Part 3: Software requirements.** This part of IEC 61508 applies to any software forming part of a safety-related system or used to develop a safety-related system.

5.2. ILO's MODEL CODE OF SAFETY REGULATIONS/LEGISLATIONS

A Tripartite Technical Conference organised by the ILO in 1948 formulated a "Model Code" of Safety Regulations for Industrial Establishments for the guidance of governments and industry. Rule 82 of this Code specifically deals with guarding machinery. According to the Code, the guards should be properly designed, constructed and used so that they will: (*a*) provide positive protection; (*b*) prevent all access to the danger zone; (*c*) cause the operator no discomfort or inconvenience; (*d*) not interfere unnecessarily with production; (*e*) operate automatically or with minimum effort; (*f*) constitute preferably a built-in feature; (*g*) be suitable for the job and the machine; (*h*) provide for machine oiling. inspection, adjustment and repair; (i) withstand long use with minimum maintenance; (*j*) resist normal wear and shock; (*k*) be durable and fire and erosion resistant; (*l*) not constitute a hazard by themselves; and (*m*) protect against unforeseen operational contingencies.

5.3. FACTORY ACT 1948

The Factories Act was passed by the Governor General of India on 23rd Sept. 1948 and it came into force on Ist April 1949. The Act was further amended in 1950, 1951, 1954 and 1976 which came into force on 26th November 1976. This act is applicable to any factory in India that employees 10 or more than 10 workers.

Aims and Objectives

(1) To provide protection to the workers employed in factories against industrial hazards and to ensure safe and better working conditions.

(2) It regulates and properly maintain various safety and welfare activities in the factories.

(3) It also regulate and maintain working hours and rest hours of workers, employment of children, women, annual leave with wages etc.

Approval, Licencing and Registration of Factories:

(1) The act specifies that before starting a factory it is necessary to take permission of state Government or chief inspector for the site on which the factory is to be situated or for the construction or extension of factory.

(2) The plans and specifications have to be approved by the factory inspector.

(3) The factory has to be registered and the license obtained after paying the necessary fees.

The application for permission should be sent to Chief Inspector, it should contain name and address of occupies, and factory, nature of manufacturing process, nature of power to be used, name of factory manager, number of workers required etc. If after submitting the application to the state Govt. or the Chief Inspector, nothing is communicated to the applicant with 3 months, the permission is deemed to have been granted.

If a state Govt. or a Chief Inspector refuses to grant permission to use a particular site, the applicant may appear to central Govt. in this connection with 30 days from the date of refusal.

The Factories Act contains specific provisions for the safety of workers. These are referred to in sections 21 to 40. They are:

(i) Fencing of Machinery

It is obligatory on the part of the management to fence machinery with guards of a substantial construction, which shall be maintained and kept in position when any part of the machinery is in motion.

(ii) Work on or Near Machinery in Motion

Any examination, adjustment or lubrication of any part of an operating machine shall be effected or carried out by a specially trained male worker wearing tight-fitting clothing. This worker, however, shall not handle a belt on a moving pulley:

(*a*) If the belt is more than fifteen centimetres in width;

(*b*) Unless the belt joint is either laced or flushed with the belt;

(*c*) Unless the pulley is normal for the purpose of a drive and not a fly-wheel or a balance wheel;

(*d*) Unless the belt, including the joint and the pulley rim, is in a state of good repair;

(*e*) Unless there is a reasonable clearance between the pulley and any fixed part of a machine or structure;

(*f*) Unless a secure foothold and, where necessary, a secure handhold are provided for the operator; and

(*g*) Unless the ladder to be used for carrying out any examination, adjustment or lubrication of any part of a machine is securely fixed or lashed or is firmly held by another person.

No woman or adolescent is allowed to clean, lubricate or adjust any part, of a machine which is in motion if it is likely to expose her or him to the risk of injury form any moving part.

(iii) Employment of Adolescents on Dangerous Machines

No adolescent shall be allowed to work on any machine which poses a danger to him unless:

(*a*) He has been fully instructed to beware of the particular danger that is likely to arise from the machine and to observe the necessary precautions; and

(*b*) He has received training on that machine or is under the supervision of a person who has a thorough knowledge of, and experience in working on, the machine.

(iv) Striking Gear or Device for Cutting off Power

In every factory, a suitable striking gear or other efficient mechanical appliance shall be provided and maintained. Driving belts, when not in use, shall not be allowed to rest or ride on a shaft in motion Suitable devices for cutting off power in an emergency shall be provided and maintained in every work room.

When a device, which is likely to be inadvertently shifted from the "off" to the "on" position, is provided in a factory to cut off power, arrangements should be make to lock it in a safe position with a view to preventing any accidental starting of the transmission machinery or any other machines to which the device is fitted.

(v) Self-Acting Machines

No transverse part of self-acting machine and no material carried there- on shall be allowed to run on its outward or inward transverse within a distance of 45 centimetres from any fixed structure which is not a part of the machine.

(vi) Prohibition of Employment of Women and Children Near Cotton-Openers

No woman or child shall be employed in any part of a factory to press cotton when a cotton-opener is at work. But if the feed-end of a cotton-opener is in a room which is separated from the delivery-end by a partition extending to the roof or to such height as the factory inspector may specify in writing, women and children may be employed in the part of the room in which the feed-end is situated.

(vii) Hoists and Lifts

In every factory, hoists and lifts shall be of good mechanical construction and of sound material; and they shall be sufficiently strong and properly maintained. Every hoist-way and liftway shall be adequately protected by a proper enclosure fitted with gates. The maximum safe working load shall be clearly indicated on every hoist or lift. A heavier load shall not be allowed to be carried on that hoist or lift.

(viii) Lifting Machines, Tackles, Chains and Ropes

In every factory, lifting machines, tackles, chains and ropes shall be of good construction and of sound material. They shall be free from defects and strong enough to carry the necessary loads.

(ix) Revolving Machinery

In every room in which grinding jobs are carried on, a notice indicating the maximum working speed of the machine shall be fixed near it.

(x) Pressure Plant

In any operation which is carried on at a pressure which is higher than the atmospheric pressure, effective measures should be taken to ensure that the safe working pressure is not exceeded.

(xi) Floors, Stairs, and Other Means of Access

All doors, steps, stairs, passages and gangways shall be of sound construction and shall be kept and maintained in a state of good repair; and they shall be free of obstructions. No substance, which is likely to cause a person to slip, shall be kept near them.

Necessary provisions should be made for a secure handhold or fencing to ensure the safety of persons working at a place from where he is likely to fall from a distance exceeding two meters.

(xii) Pits and Openings in Floor

In every factory, every fixed vessel, tank, pit or opening in the floor, which may be a source of danger, shall be securely covered or securely fenced.

(xiii) Excessive Weights

No person shall be employed in any factory to lift, carry or move any load which is so heavy as to cause him a possible injury.

(xiv) Precautions Against Dangerous Fumes

No person employed in a factory shall be allowed to enter any chamber, tank, vat, pit, flue or such other confined place in which dangerous fumes are likely to be present to such an extent as to constitute a hazard unless such chamber, tank, vat, pit or flue is provided with a manhole of a large enough size or with similar other means of egress.

(xv) Precautions Against the Use of Portable Electric Lights

In any factory:

(*a*) No portable electric light or any other electric appliance of a voltage exceeding 24 volts shall be permitted for use inside any chamber, tank, vat, pit, flue, or confined place;

(*b*) If any inflammable gas, fume or dust is likely to be present in such chamber, tank, vat, pit, flue, or confined place, no lamp or light other than the one which is flame-proof, shall be permitted to be used therein.

(xvi) Explosive or Inflammable Gas or Dust

When, in any factory, any manufacturing process produces dust, gas, fume or vapour of such nature and to such an extent that is likely to explode on ignition, all practical measures shall be taken to prevent such explosion by:

(*a*) An effective enclosure of the plant or machinery used in the process:

(*b*) The removal or prevention of accumulated dust, gas or fume; and

(*c*) The exclusion or effective enclosure of all possible sources of ignition.

(xvii) Precautions in Case of Fire

The following precautions shall be taken

(*a*) Exit doors shall not be locked or fastened and shall be capable of being easily opened; and they shall be so constructed as to open outwards;

(*b*) Proper means of escape shall be provided in every industrial establishment;

(*c*) Every door, window or other exit, through which persons can escape in the event of a fire, shall be distinctly marked in red letters in a language that is understood by workers;

(*d*) Proper arrangements shall be made to raise an alarm in the event of a fire; it would be preferable if a siren is sounded so that workers may recognise the signal as an indication that a fire has broken out somewhere in the factory premises;

(*e*) All the exits should be easily and freely accessible to all the workers in every place in the factory premises, so that they can easily make their escape when a fire breaks out; and

(*f*) All the workers shall be trained in the routine to be followed in the event of a fire in the factory premises.

(xviii) Safety of Building and Machinery

When a building or machinery poses a danger to workers, it shall not be used till it has been suitably repaired or altered.

5.4. BOILER ACT

Introduction and Scope of Act

Boiler act came in force in 1923. It insists on the registration and regular inspection of all boilers except those installed in ships or under control of railway and military. The act is applicable to all boilers, feed pipes, steam pipes and to all economizers. The act authorises the State Govt. to appoint inspectors to carry out the inspection and other provisions of the act.

Registration of Boilers

The boilers have to be registered before they can be used. The owner of the boiler has to apply for the registration. The inspector shall examine the boiler and find the maximum pressure at which the boiler may be worked. The inspector will then submit the report to the chief inspector. The owner will then get a certificate in the prescribed form, which authorizes him to use the boiler for one year at certain maximum pressure.

Certificate of Renewal

The certificates issued to the employer for using the boiler shall be renewed:

(1) After every year.

(2) If the boiler is transferred from one state to another.

(3) If some accident occurs or boiler pipes etc. do not remain in safe condition.

(4) If some alternations are done in boiler parts, etc.

Restrictions

(1) A boiler cannot be used unless it is registered, and a certificate is obtained to that effect.
(2) If it is transferred from one state to another, the owner is not allowed to use it unless it is reported to the registering authority.
(3) It cannot be used at pressures higher than that prescribed by the authority.
(4) Only a person having a competent certificate is allowed to operate the boiler.

Duties of Boiler Inspector

(*i*) The boiler inspector should see that all boilers are registered.
(*ii*) He should make sure that all boilers are working according to the provisions of the act.
(*iii*) He should check and examine boilers, their parts, mounting and accessories etc.
(*iv*) He should advise the owner of boilers in the matters of boiler maintenance, cleaning and unkeep.

Duties of Chief Inspector

(*a*) The chief inspector has to maintain records of all registered boilers.
(*b*) The boiler inspector has to examine boiler inspection reports submitted by the inspectors.
(*c*) He has to decide whether to issue the certificate for the operation of the boiler or not.
(*d*) He has to supervise and control the work of inspector.

Boiler Mounting

For efficient and safe operation of the boiler, the following boiler mountings should be fitted on the same:

(1) One blow off cock.
(2) One feed pump.
(3) One feed check valve.
(4) One fusible plug.
(5) Adequate number of manholes and sight holes.
(6) Two safety valves.
(7) One safety valve after superheater.
(8) Two water level indicators.

5.5. ELECTRICITY ACT

The electricity act was passed in 1910. The object of this act is to control and manage the electricity production, its transmission, distribution and utilization.

Supply of Energy License

A state Govt. can grant a license on application on prescribed on prescribed form and on payment of a prescribed fee to supply energy in any specified area and also to lay down electric supply for the transmission of energy.

No person other than a licensee or a person to whom sanction is granted shall transmit or use energy at a rate exceeding 250 watts and 100 volts:

(*a*) In any street, or
(*b*) In any place in which 100 or more persons are likely to be assembled or in a factory or a mine.

License should give 7 days' notice, before the commencement of transmission or uses of energy, stating his intention to the Electrical Inspector and the District Magistrate.

Purchase of Undertaking

Where a license has been granted to any person not being a local authority, the State Electricity, Board shall have the option of purchasing the undertaking after the expiry of license period. For this purpose state electricity board shall have to send an intimation to the State Govt. at least 18 months before the expiry of the period of licensee. It should also serve upon the license a notice of not less than one year requiring the licensee to see the undertaking to it at the expiry of the license period.

Central Electricity Board

A board called Central Electricity Board shall be constituted and exercise the following powers:

(*a*) To make rules to regulate the generation, transmission and use of energy.
(*b*) To provide for preparation and submission of accounts by the licensees in the prescribed form.
(*c*) To provide for securing a regular constant and sufficient supply of energy by the licensee to the consumer.
(*d*) To provide for preventing telegraph lines and magnetic observations of laboratories from being injuriously affected by generation, transmission supply or use of energy.
(*e*) To authorize any inspector to enter and inspect any place and carry out test.
(*f*) To regulate the publication of notices.
(*g*) To prescribe the qualification of electrical inspector.

Provisions of Inspection

(*i*) Any inspector or his assistant may enter or inspect and examine any place and carry out tests.

(*ii*) Inspector may serve an order to any supplier, consumer, owner or occupier, to follow out the prescribed rules.

Safety Precautions

(*i*) All electric supply lines and apparatus should be sufficient in power and size and of sufficient mechanical strength.

(*ii*) Suitable cut out should be provided in each conductor of every service line other than an earthed neutral conductor.

(*iii*) Where bare conductors are used in a building:

(*a*) The owner should ensure that they are inaccessible.

(*b*) Switches should be provided in readily accessible position, for rendering them dead whenever required.

(*iv*) The owner of every medium high and extra high voltage installation should affix permanently a caution notice on:

(*a*) Every motor, generator, transformer and other electrical plant and equipment.

(*b*) All supports of high and extra high voltage overhead lines.

Protective Equipment

The following protective equipment should be kept ready near the electrical installation for safety purpose:

(*i*) Fire buckets filed with clean and dry sand for immediate use as a means for extinguishing fires.

(*ii*) Fire extinguishers should also be provided at all generating stations, sub-stations and enclosed switch stations at convenient points.

(*iii*) First aid boxes must also be provided and be kept in charge of a person trained in first aid treatment.

(*iv*) Electrical installation work, alternations, repairs etc., should only be carried out by a Licensed Electrical Contractor and under the direct supervision of a person holding a certificate of competency issued or recognized by the State Government.

(*v*) Every installation which is connected to the supply system of the supplier, should be periodically tested and inspected at an interval at an interval not exceeding 5 years.

Supply and Use of Energy

1. Testing of Installation and Connection: On receipt of an application for a new or additional supply connections the supplier has to inspect and test the applicants installation. He may ask the applicant to make modifications in order to render the installation safe.

2. Switch Board Installations: Every switch board shall be installed as follows :

(*a*) A clear space of at least 0.914 metre in width must be left in front of the switch board.

(*b*) If the bare connections are at the back of the switch board, the space behind the board should be either less than 23 cm or more than 76 cm in width from the wall. In the latter case, there should be a passage way from either end of the switch board.

3. Variation in Frequency: The actual frequency of an a.c. supply should not vary more than 3% of the declared frequency.

4. Sealing of Meters and Cut-outs: Suppliers may seal to any cut off or meter and it is the responsibility of consumer to see that seal is not broken by any person except supplier.

5. Variation of Voltages: Actual voltage should not vary beyond 4% of the declared voltage for low and medium voltage and not beyond 12.5% for high and extra voltages.

Penalties

(*i*) A person breaking seal shall be punishable with a fine up to Rs. 200 and the consumer, when himself has not broken the seal shall be punishable with a fine up to Rs. 50.

(*ii*) Any person who violated the specified rules shall be punishable for every such act with a fine of Rs. 300.

5.6. WORKMEN'S COMPENSATION ACT

The Workmen's Compensation Act, 1923 came into force on 1 July, 1924. The act was amended in 1926 and 1929 to introduce certain changes. In 1933 a consolidating and amending act was passed to incorporate the recommendation of Royal Commission on labour. The workman's compensation act, 1976 is latest amendment which became effective from Ist Oct 1976. Main provisions of act amended in 1976 are as follows:

Workmen's Compensation act provides compensation to certain categories of workers for loss of working capacity due to accidents. This act prevents exploitation of workers in case of payment of compensation. It safeguards the workers and their families in case of death or disablement in case of workers arising from accidents. The object of awarding compensation is to replace the actual loss suffered by worker.

Application of the Act

The act applies to all workers employed in Railways, Factories, Mines, mechanically propelled vehicles, construction work and other hazardous occupations except persons employed in a clerical or administrative capacity or in Armed forces.

Employer's Liability for Compensation

If a personal injury is caused to a workman by accident in the course of his employment, the employer shall be liable to pay compensation in accordance with the provision of this act if:

(*a*) injury has been caused by accident.
(*b*) during the course of employment.
(*c*) and has resulted in workman's death, permanent or temporary disablement.

The employer is not liable to pay compensation if:

(*a*) the injury disables the workman for less than 3 days.
(*b*) the injury is caused by an accident which occurred while the workman was under the influence of drink or drugs.
(*c*) the injury is caused by wilful disobedience of the safety rules by the worker; or
(*d*) the injury is caused owing to the wilful remover or disregard of any safety guard by the workman.

But, if the worker dies of an accident due to the above reasons the employer has to pay compensation.

Amount of Compensation

Amount of compensation depends upon the following factors:

(1) Average monthly wages of the worker concerned which should not be more than Rs. 1000.
(2) The extent of injury *e.g.*
   (*a*) Death
   (*b*) Permanent total disablement.
   (*c*) Permanent partial disablement.
   (*d*) Temporary disablement.
(3) Type of work of the worker (clerical, industrial, administrative).

Notice of Accident

A worker injured in an accident should first of all give in writing a notice of the accident to the employer.

Every such notice should include:

Name and Address of the injured worker.

Cause of accident.

Date of accident.

Medical Examination

An injured worker who has submitted a notice, must present himself for the medical examination, if the employer wants. The employer must offer the medical examination free of charge and with 3 days from the time at which service of notice has been affected.

If the worker does not present himself for the medical examination then he will not have the right for compensation.

Distribution of Compensation

Payment of compensation in respect of death of workman or under legal disability shall be deposited by the employer with the commissioner. Employer should not make payment of compensation directly.

However, in case of deceased workman an employer can give an advance upto Rs. 100 to dependents of the deceased worker for meeting the immediate expenses. The commissioner, after deducting this advance, may allot the entire amount of compensations to the dependent of the deceased workman.

Occupational Diseases

If a worker suffers from any of the specified occupational disease (the list of which is given in schedule III of the act) and he worked in the factory for at least six months, then the employer has to pay compensation. Some of the occupational diseases are compressed air illness, poisoning by lead, phosphorous, mercury, arsenic, radium and other radioactive substances, x-rays etc.

Appointment of Commissioner

The State Govt. may by notification in the Official Gazette appoint any person to be commissioner for workmen's compensation for the areas specified in this modification. Every commissioner is deemed to be a public servant within the meaning of the Indian Penal Code.

IMPORTANT QUESTIONS

1. What are safety standards?
2. Discuss ISO standards for safety.
3. Discuss the various safety standards recommended in an organisation.
4. Discuss the various safety legislation adopted by the organisation.
5. What is ILO's model code of safety regulation?
6. What are objectives of factory Act?
7. Explain Boiler Act?
8. Explain Workmen's Compensation Act?

❑❑❑

CHAPTER 6

BASICS OF ENVIRONMENT

Broadly defined, the environment include all that is external to an individual-the air we breathe, the water we drink and use, the land and built structures that surrounds us—all of the natural as well as human formed conditions that influence the quality of our lines. Our interaction with the environment is continuous.

6.1. INTRODUCTION

Environment can be defined as one's surroundings, which includes everything around an organism *i.e., abiotic* (non-living) and biotic (living) environment. Abiotic environment consists of soil, water and air, while the biotic environment includes all other organisms, with which the organism comes into regular contact.

Environment may refer to a small localised area in which a specific problem is to be addressed, or it may take on global dimensions on the other hand.

The global environment consists of the lithosphere, the hydrosphere and the atmosphere. The lithosphere is the soil mental that wraps the core of the earth, in which the biological activities take place. The hydrosphere consists of the oceans, rivers, lakes, glaciers and the shallow ground water bodies that interflow with the surface water. Approximately, 70% of earth's surface is covered with water mainly in the form of oceans. The atmosphere is a blanket of gases, and suspended liquids and solids that entirely envelope the earth, extending outward several thousand kilometers.

6.2. WORKING ENVIRONMENT

The working environment refers to the environment in which a worker performs his job. It has an influence on the strain the worker acquires while performing his task and also his productivity. If the working environment is not proper then it can increase strain on the workers, reducle their efficiency and the ultimate result is low productivity.

Even the best working methods do not help to perform the tasks efficiently and effectively if the workplace where the worker works has insufficient light, high or low temp, unhygienic conditions unbearable noise, smokes and fumes, improper ventilation etc.

For better results, the working environment should be such that the workers feel comfortable and safe.

6.3. TYPES OF WORK ENVIRONMENT

Work environment may be divided into thee components: Physical, mental and social.

1. Physical Environment: It consists of physical factors example lighting, ventilation, temperature, humidity, noise, layout of machines. Poor light cause eye strain, mental fatigue, accidents and irritation. Poor ventilation and absence of fresh air make the people uncomfortable at work. The degree of moisture or humidity in the air affect speed of work. Noise cause fatigue and reduces efficiency of work. A workers efficiency will be very low if he does not have physical comforts.

2. Mental Environment: Mental environment comprises the psychological factors which influence the attitudes, behaviour and performance of a worker. It includes various instincts of the worker like fears, anger etc. Favourable mental environment should be created for the satisfaction of workers and to boost their morale.

3. Social Environment: Social environment refers to the group to which a worker belongs within a large formal group workers form small informal groups and they develop a sense of belonging to their group. The norms of the group influence the attitudes and behaviour of its individual workers. Therefore, management should attempt to create groups of likeminded people and make constructive use of groups.

6.4. NEED OF ENVIRONMENTAL CONTROL

Environmental control refers to control of working conditions surrounding the work. Control of working conditions is becoming an increasingly important responsibility of any organization. If the working conditions are unsafe or not proper it affects the health of the workers. Ill-health of the workers leads to loss of production, poor performance and reduced productive efficiency. Experience has proven conclusively that factories providing good working conditions will produce more than those maintaining poor conditions.

The economic return provided through investment in controlling the working or improving working environment is usually significant.

Control of working environment will:

(*a*) Improve worker's job performance.

(*b*) Improve productivity.

(*c*) Raise employee morale.

(*d*) Reduce absenteeism and tardiness.

(*e*) Reduce physical and mental fatigue.

(*f*) Prevention of occurrence of illness or accidents.

(*g*) Increase production.

(*h*) Improve the public relations.

6.5. EFFECT OF ENVIRONMENTAL FACTORS ON HUMAN BODY AND MIND

Environmental factors are those factors which surround the worker as he performs his task. The list of environmental factors that may have some bearing on physical well-being and therefore efficiency towards work can be pretty long. Some of the environmental factors like illumination, ventilation, temperature, noise and vibrations can be very often important factors in determining efficiency of worker because these factors may either lead to fatigue or strain or may create conditions that may accentuate the effects of fatigue, though these may not cause it. Other environmental factors like Drinking water and hygiene, Nutrition, Good housekeeping, safety, introduction of music in work environment and appropriate use of colours have relation with efficiency of work.

Noise: Noise may be defined as unwanted round. It causes fatigue and reduces efficiency of work. It is a source of disturbance to the employees and does not allow them to work with concentration. A person cannot remain indifferent to noise. Therefore, efforts should be made to minimise level of noise. There are two types of noise, namely internal and external. Internal noise is caused by converation, movement of employees through consider and going ways, running of machines etc. External noise comes from outside the factory premises. It enters through doors, windows and ventilates.

Workplace: Adequate space should be provided for workers, materials tools and equipment for their free and unhindered movement from one machine or process to another.

Ventilation: State as workplace cause discomfort and fatigue to workers and reduces their efficiency. Therefore proper ventilation is necessary to provide fresh air. Electric fans, exhaust fans and air-conditioners can be used to regulate temperature and ventilation. Electric fans, exhaust fans and air conditioners can be used to regulate temperature and ventilation. Electric fans calculate fresh air and exhaust fans expel the stale air.

❐ Illumination

One of the most important environment factors is lighting or illumination. Improper Illumination can cause accidents, discomfort to workers, defects in the eyesight of the workers in the long run. Even commonsense tells us that tasks like proof reading, watch repairing and many such other tasks which require precision in work can be handled well only if proper illumination is available. The Illumination Engineering Society (IES) has determined levels for various tasks for visual performance and visual comfort. IES points out that there are two aspects of industrial illumination: the intrinsic nature of the task and the characteristics of luminous environment. The ability to make visual discrimination is affected by the brightness contrast of details to be discriminated against the background, the size of the details to be discriminated and the time available for making discrimination. Data available on these variables, in general, indicate that for any given level of illumination, if the viewing time is decreased, the brightness contrast must be increased in order to maintain the same level of discriminability.

As the methods used in determining these variables are often too complicated and require specialised knowledge, we include here only some of the levels of illumination recommended by IES.

Recommended Standards of illumination for different types of work are given below:

	The type of work	Recommended illumination
1.	Surgical operation theatre	2500
2.	Minute and fine precision work	200–1000
3.	Proofreading	150
4.	Fine assembly work	100
6.	Fine work	70
6.	Ordinary bench work	50
7.	Normal work or auto-machine work	30
8.	Rough work	15
9.	Stair ways	10
10.	Passage ways	5

A very important factor in causing eye fatigue is glare. It may emanate from the source of light as from work area. All the glare from light source can be eliminated by the special glare Bafles devised by the RLPO (Research Laboratory of Physiological Optics). It is often believed that combination of day light and artificial light harmful for vision, but the work at RLPO has shown that it is not so, in fact, it can be often more comfortable than mere artificial light alone.

❒ Temperature

Vital proteins in the human body are adversely affected if the body is exposed to temperature exceeding 104° F. Both very high and very low temperatures make workers feel uneasy, fall ill, thereby adversely affecting their performance, efficiency and reducing both production and productivity.

The human body can fairly withstand a temperature of 104° F. The heart beat rises by 10 for a rise in body temperature of 1 °F. When the body is exposed to high temperature, large amount of persiparation process, sodium chloride is carried through the pores of the .skin and is left on the skin surface as a residue when evaporation takes place. This represents a direct loss to the system and may create a disturbance to the normal balance of fluids in the body. The result is heat fatigue and heat cramps resulting in the increase of accidents.

Loss in production also take place when working conditions are undully cold.

❒ Humidity

Humidity of the atmosphere has little effect on thermal comfort sensation at or near comfortable temperatures unless it is extremely low or extremely high. Of the various measures of humidity, the relative humidity (RH) is the most relevant, as it determines the possible evaporation rate. Moisture contents of the air or humidity functions in two adverse ways: first it prevents evaporation of perspiration, second warm humid air appears warmer than its temperature because it interferes with evaporation.

Studies have shown that between 68° and 70° humidity does not affect much the functioning of the body. But above these levels it may cause great inconvenience to the worker. Moisture from skin evaporates much more quickly in a dry than a humid atmosphere. At high temperature, skin evaporation is the most important heat dissipation channel.

❒ Vibration

When a worker operates hand-held equipment such as a chain saw or jackhammer, vibration affects hands and arms and is called hand-arm vibration exposure when a worker sits or stands on a vibrating floor or seat, the exposure affects almost the entire body and is called whole-body vibration exposure.

Vibration-induced white finger (VWF) is the most common condition among the operator is of hand in tool. Vibrations can cause changes in tendons, muscles, bones and joints and can affect the nervous system.

Whole body vibration can cause fatigue, insomnia, headache and shakiness shortly after or during exposure. The symptoms are similar to those that many people experience after a long car or boat trap. Many studies have reported decreased performance in workers exposed to whole-body vibration.

6.6. BASICS OF ENVIRONMENT DESIGN AND IMPROVED EFFICIENCY AND ACCURACY AT WORK

The indoor environment affects the health, comfort and performance of occupants. All three are important, but the cost of improved design, operation, maintenance and energy use in buildings must usually justified in terms of the expected effect on productivity.

The various environmental design factors are:

(*a*) noise
(*b*) climate (temperature humidity, ventilation)
(*c*) vibration
(*d*) lighting

(a) Noise

The most important point to be made is that the successful design is one in which noise is a consideration at every stage. A noisy site, for example, will strongly affect the overall design of a building, and may necessitate costly noise control measures. Similarly within the building the spatial separation of quiet and noisy regions will simplify the problem of achieving adequate noise insulation between them. Finally, noise insulation design must be carried through to the last detail; there is no merit, for example, in specifying a good partition and then allowing it to become punctured with service outlets, or badly fitted doors. There are some ways in which noise considerations affect the design of buildings, illustrating with three common structures: the apartment building, the office building, and the industrial plant.

❑ Apartment Building

The problem to be considered in apartment buildings and other multiple-dwelling structures is the transmission of noise from one dwelling unit to another. Noises produced in apartments vary with the occupants and with their activities of the moment. Similarly one's tolerance of extraneous noise varies with one's own activities, in fact if all the occupants of a building were always doing the same things at the same time there would be no noise problem. More commonly, however, one occupant will be trying to sleep while his neighbour watches a late television show; or a day worker will be living below someone who works in the evening and diner at 3 a.m. The successful building is one that can accommodate the wide variety of tenants and tenant activities amicably. A completely "soundproof" building is not practical; the modest objective in the following discussion is to satisfy perhaps 75% of apartment dwellers 90% of the time.

An apartment building can usually be laid out in plan so that the most critical rooms (bedrooms, living rooms) are protected from adjoining apartments by a buffer zone of non-critical areas such as bathrooms, kitchens, closets and hallways. The next best arrangement is to place quiet rooms such as bedrooms on the two sides of the party wall; in this case the separation should be a 50 db wall. Similar considerations apply to floors separating dwelling units. For separating critical areas an airborne sound transmission loss of 50 db is necessary. Since in the conventional building pattern it is not easy to vary the floor construction from critical to noncritical areas, it is usually the 50 db requirement that governs.

An additional problem of special importance in floors is impact sound (e.g. footsteps), originating as a vibration in the separating structure itself. Floors separating apartments should provide adequate insulation against impact sounds. Since a satisfactory floor design is the most costly noise control measure it is worthnoting that floor transmission is the commonest and most disturbing noise problem in existing apartment buildings.

❐ Office Building

Noise insulation requirements for an office building are not as stringent as those for an apartment building. Offices are usually occupied for only about 8 of the 24 hours, a moderate amount of business noise is usually acceptable, and interference with sleep is rarely a concern. The main requirement for partitions between tenants in an office building is speech privacy. The speech originating in one tenancy should not be intelligible in an adjoining tenancy. An exact specification depends upon the ambient noise level in the listening "room", but generally an average transmission loss of 35 to 40 db is adequate. Office building floors usually provide adequate insulation for airborne sound, and impact noise is of minor importance except when heavy machine or other sources of vibration of impact are involved.

There is the problem of noise transmission over the partition above the suspended ceiling. This may be prevented (1) by using ceiling panels backed by a heavy impervious layer that reduces sound penetration through the ceiling, or (2) by building adequate partitions in the space above the ceiling.

Executive offices and conference rooms constitute the most critical noise control problem in office buildings. Speech privacy is usually a requirement, necessitating the same considerations as above even for walls within a tenancy. An additional concern is to provide conditions quiet enough for comfortable speech over ranges of perhaps 10 to 20 ft. Meeting the speech privacy requirements will take care of most business noises except possibly business machine noise. Commonly the remaining problem is the noise produced by mechanical equipment such as ventilators. Quiet ventilator design is primarily the responsibility of the equipment manufacturer and the heating engineer.

The large general office is a compromise between sound insulation and such factors as space economy, lighting and ventilation. The main noise sources are telephone conversations, and small machines such as typewriters. A partial solution is to use sound absorbing hoods or partial partitions around the principal offenders. More mechanized equipment such as card-punching and sorting machines, and reproduction equipment should be placed in a separate room where possible. Heavy machines should be mounted on properly designed vibration-isolating bases.

❒ Industrial noise

The typical industrial plant comprises offices, factory areas, storage space, and other occupancies. The special noise control problem is the factory area. No general solution can be prescribed for factory noise since the intensity and character of industrial noises vary widely. Apart from airborne noise there is often a vibration problem, which can usually be solved by providing special foundations or vibration-isolating mountings. Sometimes the machine manufacturer is prepared to offer guidance; otherwise a qualified acoustical consultant should be retained.

The requirements are: (1) to prevent hearing impairment among machine operators, (2) to facilitate necessary speech communication among operators, and (3) to prevent the transmission of excessive noise into other parts of the building or into adjacent buildings. The building designers' share in minimizing the first two problems is to provide sound-absorbing surfaces or space absorbers within the factory space. Sound-absorbing hoods or partial enclosures around the principal noisemakers are also sometimes feasible and useful.

The third problem is dealt with as in the preceding examples, except that the frequency content of the noise should be considered in conjunction with the frequency versus transmission loss characteristics of the walls and floors used to enclose it.

(b) Climate (humidity, temperature and ventilation)

The major factors that have to be taken into account when designing buildings are the sun (solar load and solar penetration during working hours), daytime temperatures, humidity, high rainfall in some areas and prevailing breezes.

Correct orientation of buildings is essential if the sun is to be kept out of rooms and off main walls. Buildings should therefore be oriented if at all possible with their long axis on a line east-west so that the main window walls face north-south. In warm, humid climates it might be necessary to modify this orientation in order to face the buildings into the prevailing breeze. If this is done other measures such as the provision of shutters or louvers might be necessary to keep the sun out of rooms.

Roof overhangs should be large enough to keep the sun out of windows. The size of the overhang will depend on the height of the roof and the latitude of the site.

Large roof overhangs will also keep the sun off walls and thus reduce the solar load. Roofs will require insulation and/or ventilation to reduce the solar load through the roof.

The use of planting in providing shade and helping to keep buildings cool should not be forgotten. Trees can be planted adjacent to buildings to provide shade and climbing plants can be trained over verandas and roofs.

Maximum ventilation in rooms is usually required throughout the year in hot, humid climates to increase comfort. Window openings should therefore be as large as possible.

Ceilings or linings should be provided if possible to reduce heat gain from the roof and thus increase comfort.

In colder regions some form of heating will be required in winter and care must be taken to ensure the safety of the heating system. If hot water radiators and heating pipes for instance are used.

(c) Vibration

While designing environment, protection of workers from the effects of vibration should be taken care. It requires a combination of appropriate tool selection, the use of appropriate vibration-absorbing materials (gloves, for example), good work practices, and education programmes. Some examples for controlling exposure to vibration are:

❐ Anti-Vibration Tools

Using anti-vibration chain saws reduces acceleration levels by a factor of about 10. These types of chain saws must be well maintained. Maintenance must include periodic replacement of shock absorbers. A few pneumatic tool companies manufacture anti-vibration tools such as anti-vibration pneumatic chipping hammers, pavement breakers and vibration-damped pneumatic riveting guns.

❐ Anti-Vibration Gloves

Conventional protective gloves (*e.g.*, cotton, leather), commonly used by workers, do not reduce the vibration that is transferred to workers' hands when they are using vibrating tools or equipment. Anti-vibration gloves are made using a layer of viscoelastic material. Actual measurements have shown that such gloves have limited effectiveness in absorbing low-frequency vibration, the major contributor to vibration-related disorders. Therefore, they offer little protection against developing vibration-induced white finger syndrome. However, gloves do provide protection from typical industrial hazards (e.g., cuts, abrasions) and from cold temperatures that, in turn, may reduce the initial sensation of white finger attacks.

❐ Safe Work Practices

Along with using anti-vibration tools and gloves, workers can reduce the risk of hand-arm vibration syndrome (HAVS) by following work practices:

- Employ a minimum hand grip consistent with safe operation of the tool or process
- Wear sufficient clothing, including gloves, to keep warm.
- Avoid continuous exposure by taking rest periods.
- Rest the tool on the work piece whenever practical.
- Refrain from using faulty tools.
- Maintain properly sharpened cutting tools.
- Consult a doctor at the first sign of vibration disease and ask about the possibility of changing to a job with less exposure.

❐ Employee Education

Training programmes are an effective means of heightening the awareness of HAVS in the workplace. Training should include proper use and maintain vibrating tools to avoid unnecessary exposure to vibration. Vibrating machines and equipment often produce loud noise as well.

Therefore, training and education in controlling vibration should also address concerns about noise control.

❐ Whole-Body Vibration

The following precautions help to reduce whole-body vibration exposure:

- Limit the time spent by workers on a vibrating surface.
- Mechanically isolate the vibrating source or surface to reduce exposure.
- Ensure that equipment is well maintained to avoid excessive vibration.
- Install vibration damping seats.

(d) Lighting

Light has an influence on work performance. A well designed lighting scheme increases individual and group productivity and reduces accidents at work. So while designing lighting some factors should be considered:

(*i*) Industrial light should provide energy efficient illumination in quality and quantity sufficient for safety and to enhance visibility and productivity.

(*ii*) Follow standard levels of illumination to provide sufficient illumination to perform the task well without fatigue and least chance of error.

(*iii*) Combine lighting, heating, cooling and atmospheric central requirements in an integrated system with the use of higher illuminances. The lighting system provide most of energy during heating period. When cooling is required much of the lighting heat can be removed by air exhaust system.

(*iv*) It is important for lighting designer to analyze luminaire construction and reflectors finish and also to make provision for maintenance access so the system can be properly services.

(*v*) Balance all of the energy management considerations and economic factors including, initial costrs, operating and maintenance costs, versus the quantity and quality requirements for optimum visual performance. The choice of electric distribution system may affect overall economics.

(*vi*) The use of daylighting should be considered for area lighting in all industries.

(*vii*) Low ambient temp is common in frozen food plants and cold storage warehouses. Abnormally high temperature is seen in truss heights in foundries, steel mills, fuel shops etc. Equipment should be selected to operate under such condition and particular attention should be given to lamp starting and light output characteristics, if fluorescent equipment is considered.

IMPORTANT QUESTIONS

1. Define environment. What is need of environment control?
2. List the various environment factors in industry.
3. Explain the effect of temperature, illumination humidity, noise and vibrations on human body and mind.
4. Explain the types of environment.
5. Discuss basics of environment design for improving efficiency and accuracy at work.
6. Discuss the effect of lighting on working environment.

❑❑❑

CHAPTER 7

INDUSTRIAL FATIGUE

7.1. NATURE OF FATIGUE

Fatigue has been described as "negative appetite for activity". It means a reduced capacity for further work as a consequence of previous activity where a person was trying almost as hard as he could. It includes both mental and physical reactions as well as the phenomenon of boredom and monotony.

The study and control of fatigue is difficult because it has three aspects. It has been defined in terms of one of these, namely, decline in quantity of work for a given level of effort. A second aspect is a feeling of tiredness, which is not necessarily accompanied by reduced capacity to work. The third aspect is physiological change. The most prominent physiological change is accumulation of lactic acid in the blood caused by the breaking down of glycogen, or sugar, in the blood. Other physiological changes are believed to be present in the junction of the nerve and muscle cells, and under certain conditions changes in the capacity of the nerve fiber to conduct, and possibly changes in the brain. The physiological changes do not necessarily accompany changes in production as in feelings. Neither the physiologically changes nor feeling changes are as convenient to measure as are changes in production.

These three measures of activity decrement, the result of prolonged work, do not necessarily operate concurrently as to time or intensity. One can be exhibited without the other two, or all three can operate together. Physiological changes occur almost as soon as work begins, and yet they are not necessarily highly correlated with the other two decrement either in quantity or in time. There is a tendency for the feeling decrement to appear sooner than the so called work decrement. In order to effects of fatigue on production, it is necessary to control such variables as the nature, difficulty, and method of work, and incentives.

For instance, improvement in illumination and rest pauses usually raise output.

7.2. TYPES OF FATIGUE

Fatigue is of three types:

(*i*) Muscular Fatigue: It means physical tiredness due to continuous work for a certain period. It reduces capacity of body and muscles to exert.

(*ii*) Mental Fatigue: It is mental tiredness and irritation. It reduces control over thoughts and causes restlessness.

(*iii*) Nervous Fatigue: It occurs when the nerve ending in muscles fail to keep the muscles active. It is caused mainly by emotional conflict and representation of negative feelings.

7.3. MEASUREMENT OF FATIGUE

Usually only plants having some type of wage-incentive plan has accurate rate of production, hence it is mainly under this systems of wage payment that plant studies of fatigue are available. Although not the only measure, production records are the most common measure of industrial fatigue. Plants with wage-incentives are preferable for fatigue measurement not only because of presence of more accurate measures of output, but also because motivation is likely to be higher.

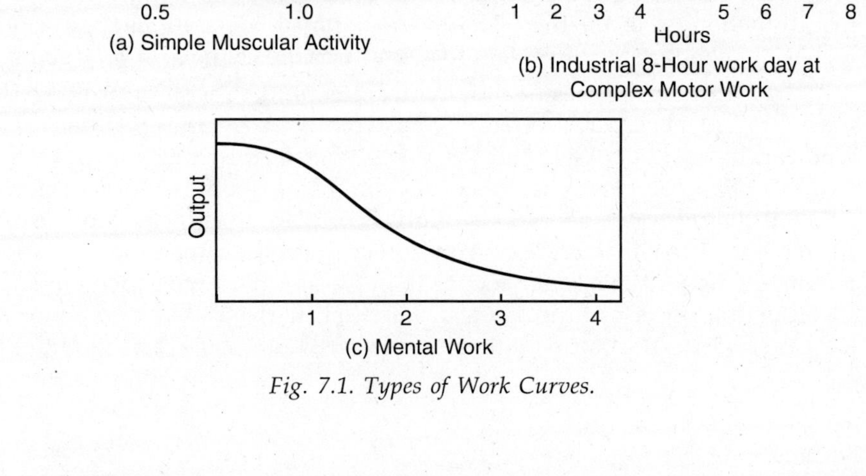

Fig. 7.1. Types of Work Curves.

A graph showing the level of performance against time spent at work is called a work curve. Since practices to alleviate fatigue depend to a large measure on the nature of work curve, its discovery and use are important. If work is uniform and successive in terms of units produced then there is no difficulty exists in obtaining such a curve. However, where work changes are frequent, in ordinary varied office work, work curves are most frequently obtained by the use of laboratory task tests. The results of these tests are then plotted and used as an index of fatigue for such jobs. Motivation might be higher in the laboratory tests than on the job.

Work curves can be typed according to the nature of task involved, namely, simple muscular, complex muscular and mental. A work curve is a composite showing the effects of fatigue and other variables over a period of time.

Simple Muscular Work

The first work curve, figure 7.1, is a simple muscular activity curve. It is derived by recording the units of output of group of muscles against the force of some weight or spring. It can be seen from the drawing that the curve for simple muscular tasks shows a short warm-up period followed by a high level of performance. A gradual tapering off then appears. Subsequently a sudden drop occurs to a point of complete exhaustion. Such work curves have a general resemblance to motor curves, but they involve simpler coordination and fewer muscles. One can conclude from a study of such curves that the more complex and rapid the task, the faster fatigue sets in. Moderate tasks permit a greater total amount of work before complete exhaustion occurs than do heavier tasks. Since each person has own work curve, it appears possible to match the workers to jobs whose energy requirements more nearly approximate his own. Discovery of a person's work curve allows the scheduling of activities and rest pauses that fit his peculiarities.

Complex Muscular Work

The second curve depicts the amounts of work done per unit of time in a typical industrial situation at a complex muscular task. This job is assumed to be motor in nature and not monotonous. The initial upward slope of curve indicates a warm-up period. This is followed by a gradual rise in production until mid morning, when a drop occurs. The afternoon curve, taken after the midday rest, is similar in appearance to the morning curve but it does not rise to as high a production peak and it falls more rapidly near the end of the day.

Mental Work

The third graph shows a typical mental work curve. As in muscular work, mental work decrement occurs in quantity, speed and accuracy of output. Again large individual differences and the nature of work tend to determine the timing and extent of such declines. The work decrement in mental tasks requiring continuous attention has been attributed to interferences know as "blocks". These intervals become lager and more frequent as the time spent at work progress. These lapses in performance are sources of mistakes. In office or industrial workplaces such blocks may cause accidents. Two experiments conducted to determine results of intense mental work suggest that a great deal of such work decline is really due to loss of interest and incentive and hot exhaustion per sec. Blocking acts as a natural way of providing rest. Mental work that is varied can consequently go on with few interruptions, but work requiring constant attention will produce many blocks and consequently a sharp drop in output with the passage of time. As shown, the mental fatigue curve generally has no warm-up period, but it shows a gradual but steady downward slope with some emphasis on decelaration at the later stages.

7.4. MITIGATION OF FATIGUE

Fatigue can be mitigated by taking following measures:

1. Fatigue may be reduced by improving environment conditions (lighting, ventilation, temp., humidity, noise).
2. A physical crowded situation results in mental fatigue and confusion. When too may workers working in small space fatigue occurs. Hence extension of working space and rearrangement of plant layout can eliminate fatigue.
3. Adequate training should be given to relieve tension and thus reducing fatigue.
4. Fatigue occurs when work is repetitive or monotonous. A change from one job to another breaks monotony and hence reduce fatigue.
5. Studies show that fatigue develops only after some hours of continuous work. Rest pauses helps to reduce fatigue to a greater extent.
6. Proper application of scheduled hours can go a long way in reducing the incidence of fatigue.
7. Regular exescise improves overall health and may decrease bedtime restlessness and hence reduce fatigue.
8. Steady, consistent bedtime routine helps to avoid fatigue.

9. Adjustment of eating and drinking habits and taking balanced diet can help to avoid fatigue.

IMPORTANT QUESTIONS

1. What is fatigue? How it is related to working environment.
2. Discuss various types of fatigue.
3. What are various methods to measure fatigue?
4. What are methods to mitigate fatigue?

CHAPTER 8

Heat Stress in Industry

8.1. HEAT STRESS

The human body is equipped to maintain an appropriate balance between the metabolic heat it produces and the environmental heat to which it is exposed. Sweating and the subsequent evaporation of sweat are the body's way of trying to maintain an acceptable temperature balance. According to Alpaugh, this balance can be expressed as a function of the various factors in the following equation.

$$H = M \pm R \pm C - E$$

where H = body heat

M = internal heat gain (metabolic)

R = radiant heat gain

C = Convection heat gain

E = Evaporation.

As long as heat gained from radiation, convection and metabolic processes does not exceed that lost through the evaporation induced by sweating, the body experiences no stress or hazard. However, when heat gain from any source is more than the body can compensate for by sweating, the result is *heat stress*.

"Heat stress is the net (overall) heat burden on the body from the combination of the body generated heat while working, environmental sources (air temperature, humidity, air movement, radiation from the sun or hot surfaces/sources) and clothing requirements.

Heat stress is a major concern of modern health and safety professionals.

Heat Stress Causes the following Illnesses

Heat edema is swelling which generally occurs among people who are not acclimatized to working in hot conditions. Swelling is often most noticeable in the ankles. Recovery occurs after a day or two in a cool environment.

Heat rashes are tiny red spots on the skin which cause a prickling sensation during heat exposure. The spots are the result of inflammation caused when the ducts of sweat glands become plugged.

Heat cramps are sharp pains in the muscles that may occur alone or be combined with one of the other heat stress disorders. The cause is salt imbalance resulting from the failure to replace salt lost with sweat. Cramps most often occur when people drink large amounts of water without sufficient salt (electrolyte) replacement.

Heat exhaustion is caused by loss of body water and salt through excessive sweating. Signs and symptoms of heat exhaustion include: heavy weakness, dizziness, visual disturbances, intense thirst, nausea, vomiting, diarrhea, muscle cramps, breathlessness, palpitations, tingling and numbness of the hands and feet. Recovery occurs after resting in a cool area and consuming cool salted drinks.

Heat syncope is heat-induced giddiness and fainting induced by temporarily insufficient flow of blood to the brain while a person is standing. It occurs mostly among unacclimatized people. It is caused by the loss of body fluids through sweating, and by lowered blood pressure due to pooling of blood in legs. Recovery is rapid after rest in a cool area.

Heat stroke and hyperpyrexia (elevated body temperature) are the most serious types of heat illnesses caused by prolonged work in hot environments. Signs of heat stroke include dry, hot skin (due to failure to sweating), body temperature often exceeding 41° C, and complete or partial loss of consciousness. The signs of heat hyperpyrexia are similar except that the skin remains moist.

8.2. PHYSIOLOGY OF HEAT REGULATION

When a person is suspected of having fever, the doctor uses a thermometer to measure the body temperature. A temperature of 98.4° F or 37° C is considered to be normal. If the temperature is above 104° F or 40° C, we get alarmed.

The temperature of different parts of the body varies. The temperature of the outer skin like hands, arms, forehead etc. is influenced by the weather conditions around. A person working near a furnace may have a warm forehead and a person working in cold may have cold hands. The temperature of the inner organs like heart, brain, liver etc. is not influenced by the outside temperature. The temperature of the inner organs have to remain in specified constant temperature called core temperature if the body has to survive. This is achieved in the body by an automatic thermal regulation process taking place within the body itself.

Thermal regulation process in the human body accurately controlled by the brain, which gets information through heat sensitive nerves located throughout the body.

Human body, like any other object loses or gains heat in three ways conduction, convection and radiation. The blood circulation regulates the heat to some degree by carrying blood from cold areas to warm areas and vice-versa. When the outer skin of the body gains heat as happens when a person works near a furnace exposed to radiant heat, secretion of sweat is increased which leads to evaporative cooling. The person feels thirsty and prompted to take water. Changes take place in heart beating and breathing and the body activities which control the generation of heat within the body. On the other hand, when the body loses heat as happens when one is exposed to cold weather, secretion of sweat is controlled and more heat is produced by muscles and internal organs. Shivering of the body in cold weather is nothing but the muscles try to generate more heat. In both cases physiological changes take place prompting the person exposed to take corrective action. In case of cold, the person feels like taking hot drinks, keeping hands, feet and body together, slapping hands and stamping feet to generate heat. In the case of hot, the person feels thirst taking cold drinks and needing rest.

If the exposure time to excess cold or hot is too long so that the regulatory mechanism within the body cannot cope up with it, the temperature of the inside organs will change which is highly dangerous.

8.3. THERMAL ENVIRONMENT

Thermal environment is related to combined effects of radiant temp., humidity air velocity etc.

The thermal environment we work in, can have a large effect on the enjoyment and effectiveness of our work. It is divided into three broad categories. The first is thermal comfort, where people feel neither too hot nor too cold, and do not perceive the temperature to be a problem. The second is thermal stress, where the thermal environment will cause clearly defined medical conditions, and can prove fatal. The third is thermal discomfort, which lie between first two. People can feel too hot or too cold, but do not display medical symptoms beyond irritability and fatigue.

8.4. MEASUREMENT OF THERMAL ENVIRONMENT

Feeling of hot or cold depends on

- air temperature
- relative humidity of air
- presence of hot or cold object in the surrounding area
- Presence of air movement (breeze, ventilation)

Various methods of measuring occupational heat exposure combine these environmental factors to obtain a single number as a measure of overall heat load. The most commonly used measure in the workplace is the wet bulb globe temperature (WBGT) index.

The Wet Bulb Globe Temperature (WBGT)

The wet bulb globe temperature is calculated using a formula that takes into account air temperature, speed of air movement, radiant heat from hot objects, sunshine and body cooling due to sweat evaporation.

How to Calculate the WBGT Index?

The wet bulb globe temperature (WBGT) is calculated by using the following equations.

- For outdoors **with** direct sun exposure:

$$\text{WBGT} = 0.7 \times \text{Temp}_{\text{wet}}\ \text{bulb} + 0.2 \times \text{Temp}_{\text{globe}} + 0.1 \times \text{Temp}_{\text{air}}$$

- For indoors or outdoors **without** direct sun exposure:

$$\text{WBGT} = 0.7 \times \text{Temp}_{\text{wet}}\ \text{bulb} + 0.3 \times \text{Temp}_{\text{globe}}$$

where

$\text{Temp}_{\text{wet bulb}}$ natural wet bulb temperature measured by using a thermometer whose bulb is covered with wet cotton cloth and is cooled by the natural air movement.

$\text{Temp}_{\text{globe}}$ temperature measured using a black globe thermometer.

Temp_{air} temperature measured using a conventional thermometer.

All temperatures are to be expressed in °C.

8.5. THERMAL COMFORT

Thermal comfort is that condition of mind which expresses satisfaction with the **thermal environment**. Thermal environment is those characteristic of environment which affects a person's heat loss. In terms of bodily sensations, thermal comfort is a sensation of *hot, warm, slightly warmer, neutral, slightly cooler, cool and cold*.

Thermal comfort in the workplace is dependent on the combination of environmental and personal factors. **Environmental factors include**:

- o air temperature
- o humidity
- o radiant temperature
- o air movement
- o level of activity
- o clothing worn

Personal factors include:

- o perception
- o physical fitness
- o medications taken
- o the body fluid and salt balance
- o acclimatisation.

8.6. THERMAL COMFORT INDICES

Over the past 50 years much research effort has been devoted to developing indices predicting thermal comfort. 3 main indices are currently used.

(a) Fanger's Comfort Equation

Fanger's basic assumption—thermal comfort is defined in terms of the physical state of the body rather than that of the environment *i.e.* what we actually sense is skin temperature and not air temperature. For thermal comfort need:

- Thermal balance—rate of heat loss = rate of heat production. This is a necessary but not sufficient condition for comfort *e.g.* sweating may lead to heat balance but may not be comfortable.
- Mean skin temperature —should be at appropriate level for comfort. (NB Skin temperature for comfort decreases with increased activity)
- Sweating—comfort is a function of a preferred sweating rate, which is also a function of activity and metabolic rate.

Fanger derived his comfort equation from extensive survey of literature on experiments, on thermal comfort.

Fanger's Comfort Equation. This equation contains terms which relate to:

- **Functions of Clothing:**
  - o Icl = clothing insulation in clothes
  - o fcr = ratio of clothed/nude surface area

- **Functions of activity:**
  - o H = Metabolic heat production (w/m^2)
  - o M = Metabolic free energy production (external work) ($w\backslash m^2$)
- **Environmental variables:**
  - o Ta = Air temperature (°C)
  - o Tr = Mean radiant temperature (°C)
  - o v = Relative air speed (m/s)
  - o Pa =Vapour pressure of water vapour (mb)

    hc = convective transfer coefficient (w/m^2 K)

H – 0.31(57.4 – 0.07 H – Pa) – 0.42(H – 58) – 0.0017M(58.7 – Pa) – 0.0014M (34 – Ta) = 3.9 × 10–8fcl {Tcl + 273)4 – (Tr + 273)4} + fcl hc (Tcl – Ta)

where the clothing surface temperature, Tcl, is given by

Tcl = 35.7 – 0.0275H + 0.155Iclo {H – 0.31(57.4 – 0.07H – Pa) – 0.42(H – 58) – 0.0017M(58.7 – Pa) – 0.0014M(34 – Ta)}.

In addition to this, discomfort occur when the skin is wetted (sweat, water, etc.) and Fanger has produced an equation for skin wettedness which can be used as a test to exclude conditions which satisfy the comfort equation:

(b) Predicted Mean Vote (PMV)

The problem with Fanger's equation is that when people are not satisfied, this is not a measure of how uncomfortable deviation is; therefore Fanger developed PMV = mean vote on ASHRAE scale (Hot warm slightly warm neutral slightly cool cold). PMV can be predicted from Fanger's equation thus:

PMV = 4 + (0.303 exp (–0.036H) + 0.0275) × {8.57 + 0.46H + 0.31Pa + 0.0017HPa + 0.0014HTa – 4.13 fcl (1 + 0.001dT) (Tcl – Tr) – hcfcl (Tcl – Ta)}

where Tcl (surface temperature of clothed body)

= 35.7 – 0.0275H + 0.155 Iclofcl (4.13 (1 + 0.Old Temp) 1 + 0.155 Iclofcl (4.13 (1 + 0.Old Temp) thc

where hc = 2.4 (Tcl – Ta)0.025 or 12.1 square root of v (air speed) which is greater and dT = Tr – 22.

Predicted Percentage of Dissatisfied (PPD) can be derived from PMV and this relates to the temperature range.

While PMV gives good results for standard conditions of sedentary activity and light clothing, it has yet to be validated across a range of clothing and activity. Also PMV is effectively a measure of the thermal load of the thermoregulatory system, therefore a comfortable person who increases his metabolic rate by say 20 w/m^2 (0.34 met) will experience the same change in thermal load whatever his clothing insulation BUT with high clo values the resulting increasing skin and body temperature must be greater than for low clo values therefore one intuitively should expect a greater change in thermal sensation.

In experimental tests of thermal comfort Fanger's equation has proved very successful. The experiments all show that comfort relates to perceived skin temperature and not to environmental variables or clothing. Experiments also show that for sedentary work and light clothing led to a preferred temperature close to the 25.6°C predicted by Fanger's equation. Also to date no evidence has been found for systematic individual differences in preferred temperatures. There are no effects of age, sex, race, etc.; differences in preferred temperatures often due to clothing differences.

(c) Skin Wettedness Equation

$$w = (H - 58) / [4.6 \text{ hc } (57.4 - 0.07H - pa)] + 0.06$$

when w is too high, leading to discomfort.

However, upper limit of w depends on metabolic rate and therefore limit is estimated using

$$w \; 0.0012M + 0.15$$

8.7. THERMAL LIMITS FOR COMFORT

Comfortable limits of temperatures in which a person feels comfortable is far below the temperatures he can stand without damage and ill effects on body functions. The feeling of comfort, to some extent is subjective and varies from person to person. A person used to cold climate may find a room temperature of 25° C uncomfortably hot while another person coming from a tropical country might consider 25° C as too cold to bear. A person sitting and doing a clerial job will feel more cold than another person doing heavy manual job, in the same atmospheric conditions. The type of clothing worn also affects the feeling comfort.

Feeling of comfort is not decided by temperature alone. It is also dependent on:

- humidity
- air movement
- radiant heat

For this reason, temperature conditions in living and working atmospheres are expressed in terms of "Effective Temperature" a scale or measure which takes into account all three factors. Effective temperature can be read from charts once air velocity, humidity and radiant heat are known.

Another measure which takes into account all the three factors is the WBGT index (This is explained above). A WBGT index of 79° F is considered a comfortable level for most people.

8.8. CONTROL OF HEAT EXPOSURE

Exposure to high temperature and heat stress is possible in many situations in industry. Some examples are:

(*i*) Working near or around boilers, steampipes and hot processes.

(*ii*) Working near furnaces, heat treatment tanks etc.

(*iii*) Working in foundries and forge shops.

(*iv*) Working on operations like welding, gas cutting.

(*v*) Working in hot confined spaces.

(*vi*) Working in very hot weather.

(*vii*) Working in direct sunlight in summer.

The way to tackle this problem should begin with the identification of source of heat and an evaluation of temperature conditions by actual measurements (temperature humidity and air velocities). This should be followed by assess the effect of heat on persons exposed. These will give an idea of the extent and magnitude of the problem.

The following measures can be taken to control heat exposure in industry.

1. **Control at Source:** Separation of heat sources from general working area can prevent exposure to heat to many persons. For example, a furnace emitting heat in a machine shop can be shifted to a room enclosed from all four sides by a brick wall. This will prevent all persons other than furnace operator from being exposed to heat.
2. **Insulation:** Insulating steam pipes, hot processes and hot surfaces can stop the atmospheric air from getting heated by them.
3. **Control of Radiant Heat:** Radiation shielding can be employed in case of heat stress from radiant heat. For example in case of radiant heat from a furnace, a shiny aluminium sheet vertically placed between operator and furnace will reflect the heat back and protect the operator. Reflection of heat from walls and ceilings can also be reduced by shielding.

4. **Ventilation and Air-Conditioning:** Ventilation, localized air-conditioning, and cooled observation booths are commonly used to provide cool work stations. Cooled observation booths allow workers to cool down after brief periods of intense heat exposure while still allowing them to monitor equipment.
5. **Reducing Metabolic Heat Production:** Automation and mechanization of tasks minimize the need for heavy physical work and the resulting buildup of body heat.
6. **Adjusting humidity and air Velocity:** Often comfortable levels can be obtained by adjusting humidity and air velocity. In dry hot climates, humidification will bring about comfort. In humid and hot climate, augmentation of air flow like use of circulators can reduce the discomfort. However these changes should be made only after a scientific evaluation of prevailing conditions.
7. **Reducing Exposure Time:** In cases where the improvement of conditions is not possible, the persons can be protected from the exposure of heat by local relief. One way is to reduce the exposure time. In case, only some areas in the factory are hot, it might be convenient to stagger the work such that a person works for some time in hot area and then goes to a colder area, for the remaining time.
8. **Rest pauses:** Introducing rest pauses in between work periods. The rest has to be in a cool room and the persons should really rest. The period and frequency of rest pauses have to be decided on the basis of actual assessment of heat stress by measurements. Rest pauses allow the persons to recuperate and bring back heart beats, blood pressure and circulation to normal.
9. **Balanced Food:** Working in hot climate, puts more demands on the body systems in terms of energy output. Naturally, the energy has to come from the food taken in. So the balanced food which can give necessary calories, is important. Periodic medical examination is also recommended.
10. **Beverages:** Persons engaged in work in hot climates should maintain the water balance and salt balance in the body. Drinking plenty of water and taking salt tablets are also recommended. Quantity of water and salt tablets to be taken is best decided in consultation with a doctor or the plant medical officer.
11. **Personal Protection:** Ordinary clothing provides some protection from heat radiated by surrounding hot surfaces. Specially designed heat-protective clothing is available for working in extremely hot

conditions. In hot and humid workplaces, light clothing allows maximum skin exposure and efficient body cooling by sweat evaporation.

Workers who move back and forth between very hot, dry indoor environments and cold winter outdoor environments find that long underwear moderates the extremes in temperatures.

Eye protection which absorbs radiation is needed when the work involves very hot objects, such as molten metals and hot ovens.

Work that requires the wearing of impermeable clothing presents an added heat burden as the clothing reduces the body's ability to dissipate heat. Under such circumstances, it is often necessary to reduce the exposure limit values of WBGT to levels below those appropriate for workers light clothing.

IMPORTANT QUESTIONS

1. Give the formulation for heat load requirement of industrial buildings.
2. What is thermal environment and measurement method?
3. Define thermal comfort. What are thermal limits for comfort? Explain indices of thermal comfort of heat stress.
4. Explain heat stress and its effects on the efficiency of workers.
5. Explain physiology of heat regulation.
6. What are various measures taken to control heat exposure in industry?

CHAPTER 9

VENTILATION AND AIR-CONDITIONING

9.1. PURPOSE OF VENTILATION

The purpose of ventilation is to maintain the right exchange of gases in the body. This occurs by supplying tissues with oxygen (O_2), a gas needed for metabolic processes and eliminating carbon dioxide (CO_2) which accumulates as a result of processes generating energy.

Ventilation becomes necessary where people stay because human body consumes oxygen and releases carbon dioxide. In a closed room without ventilation the oxygen content in the breathing air will diminish and cause discomfort. The contamination of air by sweating and body odour also necessitates ventilation. Human body dissipates heat which if not transferred to the atmosphere will result in temperature rise of the room and consequent discomfort.

In factories many persons working together itself causes air pollution. Apart from the human body, machines and processes contribute to heat dust, gases and fumes generated in the process also make the work environment more uncomfortable and sometimes dangerous.

So the purpose of ventilation is to improve the quality of air around the workers by adequate supply of oxygen or air required for respiration. Also to provide thermal comfort by maintaining adequate heat balance of the body and to get rid of contaminations produced in the various processes of the industry concerned.

9.2. TYPES OF VENTILATION

Ventilation is done by replacing (in full or part) the contaminated or foul air by fresh air from the atmosphere outside. Outside atmosphere, of course, has an unending supply of fresh, good quality air (except in unnatural cases).

The replacement of foul air by fresh air is achieved by **Natural General Ventilation, Mechanical General Ventilation, Dilution Ventilation, Local Exhaust Ventilation.**

Natural Ventilation

Natural Ventilation can be effected by wind action and stack effect. It is achieved by providing enough openings at convenient locations so that natural wind can pass through the space and automatically replace the foul air. To take the proper advantage of wind action, in the design stage itself of the plant, these aspects should be looked into. The distribution of windows, doors and openings, their number, the ceil height, as well as the orientation of the structure (so as to avail the prevailing winds) need careful consideration. In case of an enclosure, when the air inside and outside the enclosure has different temperatures, ventilation may be effected by stack effect. Openings near the roof like ventilators and other vents can be provided for stack effects. In this case, the cool outside air enters the enclosure through openings at lower level (like windows) whereas the warm air in the enclosure will pass through openings near the roof or ceiling levels. It will be necessary in case of natural ventilation, to take into account the position of buildings with respect to wind direction prevailing, position of sun and admission of sunlight, outside temperature during different parts of the year and the desired inside temperature conditions.

This is a cheap and effective method in cases when

- there are no external contaminants generated (like dust, gases etc.)
- outside atmosphere air itself is not contaminated (*e.g.*, dust storm, too cold)

Special types of ventilators which increase the natural flow of air are also commercially available and can be used to augment natural ventilation.

Mechanical General Ventilation

Mechanical general ventilation is done in two ways:

- Blowing, air form outside into the room resulting in the development of a slight positive pressure inside and constant flow of air from outside to inside and then out.
- Sucking out the air inside the room with the help of exhaust fans and developing negative pressure inside the room. This will draw outside air inside to replace the foul air thrown out by the fan.

Both the methods have their own advantages and areas of specific applications. For instance, a rest room for workers is provided adjacent to the spinning department of a rayon factory where hydrogen sulphide may be present.

In the spinning room, it is advisable to create negative pressures so that air from outside is sucked inside. But in the rest room, it will be desirable to blow air from outside as creating negative pressures will result in part of the foul air from the spinning section also entering the rest room. Fresh air from outside comes to rest room and a part of it escaping to the spinning shop can cause to harm. In the case of positive pressure systems, it will be possible to clean the air before blowing inside which is not easy in the case of negative pressure.

Mechanical ventilation can be effected as follows:

- By using exhaust fans.
- By the use of positive ventilation i.e., by the use of supply fans coupled with suitable ducts to force fresh air from outside the enclosure.
- By combination of exhaust fans and positive ventilation.

Dilution Ventilation

The amount of uncontaminated outside air introduced mixes with the inside air and serves to dilute and reduce the concentration of air contaminants to acceptable level to which a worker can be safely exposed for eight hours a day.

Dilution ventilation is usually applied to the control of contaminants in situation meeting the criteria:

- To control vapour (+ organic) from low toxicity solvents.
- To control contaminants released over such a large area or in such a manner that exhaust ventilation is impossible, impractical or expensive.
- Small quantities of contaminants released into the work room.
- Rate of contaminant release should be reasonably constant to avoid inadequate dilution during periods of peak contaminant release.
- No corrosion or other problems from the diluted contaminants in the work room air
- Sufficient distance from the worker to the contaminants source to allow dilution to safe levels.

  Dilution Ventilation is Prohibited: Control emission of very toxic air contaminants example formaldehyde or other carcinogenic chemicals. For effective dilution ventilation, the exhaust outlet and air supply must be so located that all the air employed in the ventilation passes through the zone of contamination and avoid re-entrance of the exhausted air.

Advantages and Disadvantages of dilution ventilation

Advantage	Disadvantage
1. Simplicity	1. Large volumes of dilution air needed
2. Low original cost	2. Employee exposures are difficult to control near source

Local Exhaust Ventilation

It is used to capture or contain contaminants released at their source before escape into the work room environment. It removes contaminants rather than just dilute them.

A typical local exhaust system will consist of:

- A "hood" to capture and direct the contaminants into pipeline.
- A pipeline (or ducting) to transport contaminants.
- A suitable fan to generate sufficient air velocity to capture and transport the contaminants.
- Collectors/filters to separate the contaminant from air.

Hood

A pipe terminating near the source of contaminants is the simplest exhaust hood. It is not effective because it allows air from all directions to be sucked in resulting only a portion of the contaminants is removed and a lot of clean air around is also sucked into the pipeline.

Providing flange around the opening reduces suction from one side of pipe, thus increasing suction from front. Thus the effectiveness is increased. This arrangement finds practical application in hoods for grinding machines and exhaust hoods for welding.

A cannoy type hood is a improved version and is more effective. The side draught exhaust hood is also effective and has the advantage that operator has freedom of movement. The factors that are taken into account for the correct selection and design of exhaust hood are:

- The shape of hood.
- Distance of the hood from the source and the directions from which contaminants are released.
- The area of opening necessary, which depends upon the job itself.
- Flow of air inside the hood and the resistance to air flow offered by the hood.
- The capture velocity required – i.e. the velocity of air necessary to capture the particles effectively.
- A poorly designed hood will result in complete failure of system irrespective of the capacity of fan used.

The Pipeline (Ducting)

The use of pipelines in exhaust systems is to transport the contaminants trapped by the hood to the fan and then to the exhaust. It could by a simple pipe connecting the exhaust fan and one hood, or a complicated network of pipelines connecting many hords. The important considerations are:

- Transport velocity required and the quantity of air to be handled.
- Corrosive, abrasive or sticking properties of the contaminants being transported.
- The length of pipeline required.
- Bends, joints and diversions required to be included—which increase the resistance to flow.

Fan

The fan is used to produce sufficient air velocity to capture and transport the contaminants. The fan selected should have sufficient capacity to handle the required quantity of air and provide transport velocity needed and also the power to overcome resistance of system starting from hood to exhaust. The types of fans used commonly are:

- ❒ Axial flow fans with or without guidevanes which are suitable for low pressure atmospheric applications.
- ❒ Centrifugal fans with:
  - forward curved blades for low and medium pressure applications.
  - aerofoil blades.
  - backward curved blades for medium and high pressure systems and high air velocities.
- ❒ Propeller fans for low pressure systems and atmospheric applications like exhaust from a room.

Collectors

Apart from entrapping and conveying the contaminants, it becomes sometimes necessary to separate and collect the particles captured. The purpose is either to prevent contaminants from re-entering into the atmosphere or to reuse the material in the process and thus avoid wastage. The requirements of good collector are:

- Cleaning efficiency should not come down with time.
- Low cost and durability.
- No operational difficulties in terms of maintenance and cleaning.

There are various types of collectors available, the main ones being

(*i*) Gravity Chambers

Where dust particles are substantially heavier than air, they are allowed fall to a chamber by their own weight, and get collected in a tray. The inlets and outlets are so spaced that the maximum quantity gets collected.

(*ii*) Centrifugal or Inertia Separators

Here the air inside the collector takes a circular path and due to high velocity, the centrifugal force throws the particles away from the flow which are subsequently deflected and guided to a tray.

(*iii*) Filter

Filters interposed in the airflow traps the contaminants and releases clean air.

(*iv*) Electrostatic Precipatators

A plate or tube charged with electricity installed in the air path attracts dust particles and releases only clean air.

(*v*) Wet Scrubbers

Water is sprayed at or through the air path which wets the contaminants, and collects them.

Use of Local Exhaust Ventilation

1 The most effective means of controlling air contaminants is to capture and remove the air contaminents at their source with LEV and to prevent them from being carried away by air currents into the breathing zones of the worker.

2. Major release (toxic) of localized sources of contaminants.

9.3. VENTILATOR

According to the way of functioning, ventilators may be classified into devices which act on the body (negative pressure body ventilators and other body ventilators) and devices which forces intermittent pressure to the airways (volume-cycled and pressure-cycled ventilators).

Body ventilators include (*i*) devices which assist respiratory muscles by intermittently lowering pressure around the chest and abdomen so that air enters the nose and mouth and (*ii*) devices which apply forces directly to the body to displace respiratory muscles mechanically. Iron lung, chest shell-style ventilators, wrap-style ventilators, oscillator, rocking-bed, exsibblation belt are some type of body ventilators.

Volume-cycled ventilators: are very reliable devices. For this reason they are used in severe respiratory insufficiency which leads often to tra cheostomy. It is advisable to choose portable ventilators – easy to fit to the wheelchair with back-up battery and alarm for possible defaults of the machine.

Pressure-cycled ventilators: These devices deliver either a continuous flow of air into the airways via the nose (CPAP) or one can adjust the positive pressure delivered durings inspiration separately from those delivered during expiration, the flow of air being still continuous (BiPAP).

In comparison with volume-cycled ventilators pressure-cycled ventilators are cheaper, are easier to use and are lighter.

9.4. AIR-CONDITIONING

The control of the quality, quantity, temperature and humidity of the air in an interior space is called air-conditioning

An air-conditioner is an assembly of equipment for the simultaneous control of air temperature, relative humidity, purity, and motion.

9.5. WORKING OF AIR-CONDITIONERS

The Basic Idea

An airconditioner is basically a refrigerator without the insulated box. It uses the evaporation of a **refrigerant**, like Freon, to provide cooling. The mechanics of the **Freon** evaporation cycle is the same in a refrigerator as in an air conditioner (the term Freon is generically "used for any of various nonflammable fluorocarbons used as refrigerants and as propellants for aerosols).

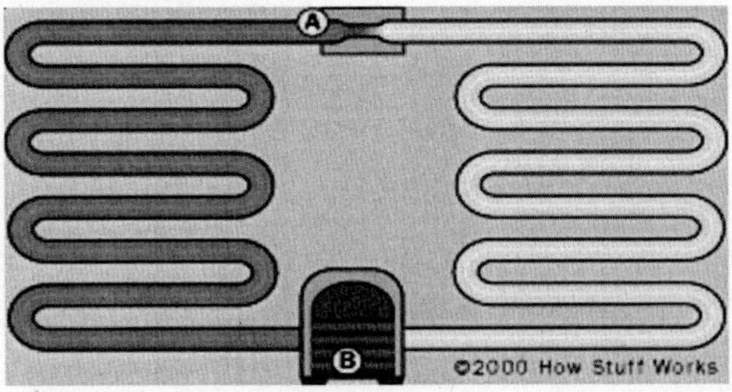

Fig. 9.1.

Diagram of a Typical Air-Conditioner

This is how the evaporation cycle in an air-conditioner works:

1. The compressor compresses **cool Freon gas**, causing it to become **hot, high-pressure Freon gas**.
2. This hot gas runs through a set of coils so it can dissipate its heat, and it condenses into a **liquid**.
3. The Freon liquid runs through an expansion valve, and in the process it evaporates to become **cold, low-pressure Freon gas**.
4. This cold gas runs through a set of coils that allow the gas to absorb heat and cool down the air inside the building.

Mixed in with the Freon is a small amount of a lightweight oil.

9.6. TYPES OF AIR-CONDITIONING SYSTEMS

Window Units

A **window air-conditioner** unit implements a complete air-conditioner in a small space. The units are made small enough to fit into a standard window frame. You close the window down on the unit, plug the unit in and turn it on to get cool air. If you take the cover off of an unplugged window unit, you will find that it contains:

- A **compressor**
- An **expansion valve**
- A **hot coil** (on the outside)
- A **chilled coil** (on the inside)
- **Two fans**
- A **control unit**

The fans blow air over the coils to improve their ability to dissipate heat (to the outside air) and cold (to the room being cooled).

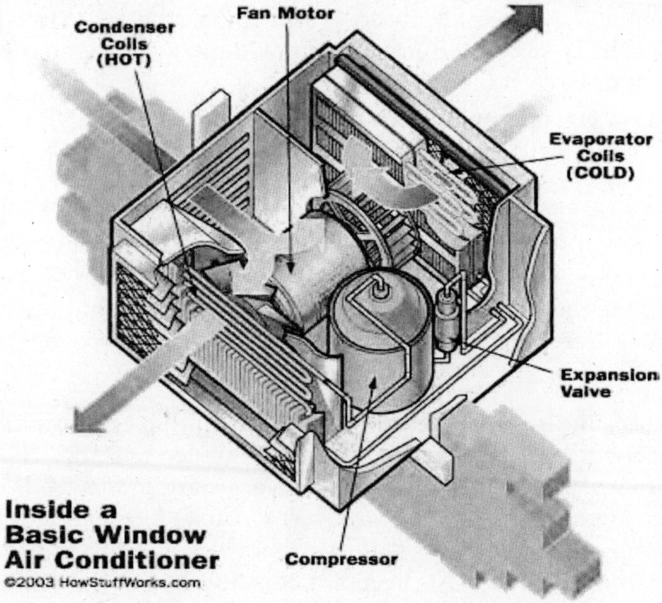

Fig. 9.2. Inside a Basic Window Air-Conditioner

Split-system Units

A **split-system air-conditioner** splits the hot side from the cold side of the system, like shown in the figure 9.3.

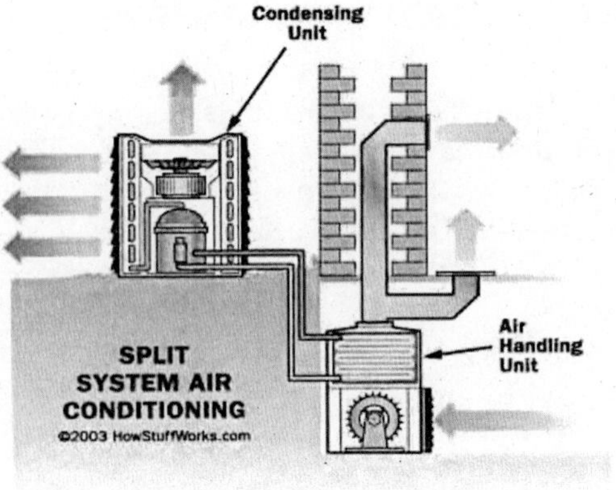

Fig. 9.3. Split System and Conditioning

The **cold side**, consisting of the expansion valve and the cold coil, is generally placed into a furnace or some other **air handler**. The air handler blows air through the coil and routes the air throughout the building using a series of ducts. The **hot side**, known as the **condensing unit**, lives outside the building.

The unit consists of a **long, spiral coil** shaped like a cylinder. Inside the coil is a **fan**, to blow air through the coil, along with a **weather-resistant compressor** and some **control logic**. This approach has evolved over the years because it is low-cost, and also because it normally results in reduced noise inside the house (at the expense of increased noise outside the house). Besides the fact that the hot and cold sides are split apart and the capacity is higher (making the coils and compressor larger), there is no difference between a split-system and a window air-conditioner.

Chilled-water System

In large buildings and particularly in multistorey buildings, the split-system approach begins to run into problems. Either running the pipe between the condenser and the air handler exceeds distance limitations (runs that are too long start to cause lubrication difficulties in the compressor), or the amount of duct work and the length of ducts becomes unmanageable. At this point, it is time to think about a **chilled-water system**.

In a chilled-water system, the entire air-conditioner lives on the roof or behind the building. It cools water to between 40 and 45 F (4.4 and 9.2 C). This chilled-water is then piped throughout the building and connected to air handlers as needed. There is no practical limit to the length of a chilled-water pipe if it is well-insulated.

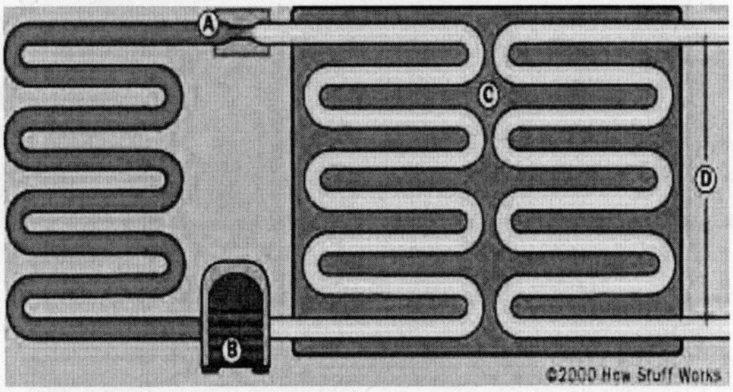

Fig. 9.4.

9.7. UNITS

BTU and EER

Most air-conditioners have their capacity rated in **British thermal units (BTU)**. Generally speaking, a BTU is the amount of heat required to raise the temperature of one pound (0.45 kg) of water 1 degree Fahrenheit (0.56 degrees Celsius). Specifically, 1 BTU equals 1,055 joules. In heating and cooling terms, 1 "ton" equals 12,000 BTU.

A typical window air-conditioner might be rated at 10,000 BTU. For comparison, a typical 2,000-square-feet (185.8 m^2) house might have a 5-ton (60,000-BTU) air-conditioning system, implying that you might need perhaps 30 BTU per square foot. (Keep in mind that these are rough estimates. To size an air conditioner for your specific needs, contact an HVAC contractor.)

The **energy efficiency rating (EER)** of an air-conditioner is its BTU rating over its wattage. For example, if a 10,000-BTU air-conditioner consumes 1,200 watts, its EER is 8.3 (10,000 BTU/1,200 watts). Obviously, you would like the EER to be as high as possible, but normally a higher EER is accompanied by a higher price.

✎ IMPORTANT QUESTIONS

1. State and explain objectives of ventilation.
2. Differentiate between natural ventilation and mechanical ventilation.
3. Explain Dilution Ventilation.
4. Explain Local Exhaust Ventilation.
5. What is air-conditioning ? How air-conditioners work?
6. What are various types of air-conditioning systems?

□□□

CHAPTER 10

NOISE

10.1. INTRODUCTION

The ability to make and detect sound provides human beings with the facility to communicate each other as well as receive useful information from the surroundings. Sound can provide the warning (the first alarm), useful information (a whistling tea kettle) and enjoyment. A low sound is pleasant whereas a loud sound is unpleasant and is commonly referred to as "noise".

Noise can be defined as an unpleasant and unwanted sound that interferes significantly with the comfort, health, or welfare of persons, or with the full use of enjoyment of property.

10.2. TYPES OF NOISE

Impulse or Impact Noise

Noise consisting of random occurrences of energy spikes having random amplitude and spectral content. Impulse noise is noise that is caused by a high amplitude, short-duration noise pulse.

Examples of impulse noise are thumps, scratches (including ticks and clicks) and crackle. Of all types of impulse noise, crackle, which sounds like bacon frying can be the most difficult to remove.

Continuous or Steady Noise

Noise with negligible small fluctuation levels within the period of observation is called continuous noise. Continuous noise of excessive level can damage the ear and such situations should be carefully avoided.

The ear can apparently be trained to resist the damaging effects of continuous noise, within limits of course.

Intermittent or Fluctuating Noise

Noise whose level varies continuously and to an appreciable extent during the period of observation. It consists of wide mixture of frequencies.

While continuous noise can cause auditory damage, the threshold level can be considerably higher than currently assumed. Most dangerous for the ear are powerful impulses.

Whether a given sound is as pleasant as music or as unpleasant as noise depends on loudness, duration, rhythm and the mood of the person. But loudness is definitely the most significant criterion which converts sound into noise. Exposure to loud noise is indeed annoying and harmful too.

Noise is a physical form of pollution and is not directly harmful to the life supporting systems namely air, soil and water. It's effects are more directly on the receiver *i.e.,* man. Noise pollution is the result of modern industrialized urban life and cogestion due to over population. Even though noise pollution is not fatal to human life, yet its improvement cannot be overlooked because repeated exposure to noise reduce the sleeping hours and productivity or efficiency of human being. It also effects the peace of mind and invades the privacy of a human being.

10.3. UNITS OF NOISE

The decibel (db) is used in environmental noise pollution as a measure of sound intensity level. It is defined as db = 10 $\log_{10}$ (I/I_o)

where 'I' is sound intensity in watt/m^2 and 'Io' is reference sound intensity (approx. 10^{-12} watt/m^2)

Decibel (db) can also be used as a measure of sound pressure level, in that case it is defined as,

$$\text{db} = 20 \log_{10} \left(\frac{P}{P_0} \right)$$

where 'p' is root mean square sound pressure in Pascals (N/m^2) and 'Po' is reference rms sound pressure

approx 2×10^{-5} Pa (or N/m^2)

For all practical purposes (*i.e.,* the standard atmospheric conditions), sound intensity level and sound pressure level are equal in magnitude to each other.

The db scale beings from zero, which represents the faintest sound which is audible to a normal ear in the decibel scale, each ten fold increase is represented by 10 db e.g.

Faintest sound is 0 dB,

$$\therefore \quad \text{db} = 10 \log (I_0/I_0) = 10 \times 0 = 0$$

Ten times more intense sound is 10 db,

$$\because \qquad \text{db} = 10 \log \left(\frac{10 I_0}{I_0}\right) = 10 \times 1 = 10$$

Hundred times more intence sound is 20 db,

$$\because \qquad \text{db} = 10 \log \left(\frac{100 I_0}{I_0}\right) = 10 \times 2 = 20$$

Thousand times more intense sound is 30 db

$$\because \qquad \text{db} = 10 \log \left(\frac{1000 I_0}{I_0}\right) = 10 \times 3 = 30$$

and so on

The sound intensity level increases by 3 db, if the sound intensity doubles likewise, if the sound intensity level decreases by 3 db, the sound intensity is halved.

If I = 10 I_o; then db = 10 log $\left(\frac{10 I_o}{I_o}\right)$ = 10 × 1 = 10

Now if I = 2 × (10 I_o); then db = 10 log $\left(\frac{20 I_o}{I_o}\right)$ = 10 × 1.3 ≈ 13

Further, if I = 2 × (2 × 10 I_o), then

$$\text{db} = 10 \log \left(\frac{40 I_o}{I_o}\right) = 10 \times 1.6 \approx 16$$

and so on

The procedure of adding decibels is quite complicated. The procedure can be simplified by a great deal by using the graph shown.

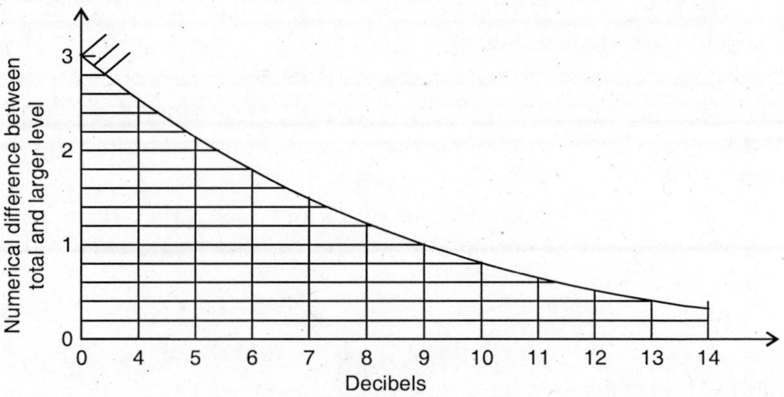

Fig. 10.1.

As a rule of thumb, adding two equal sound increases the SPL (sound pressure level) by 3 db, and if one sound is more than 10 db louder than a second sound, the contribution of later is negligible.

The chart shows combining different sound pressure level. For example, combine 60 and 50 db. The difference is 5 db. The 5 db line intersect the curved line at 1.2 db, thus the total value is 61.2 db.

The covering of a sound with a louder one is known as **masking.** Speech can be masked by industrial noise as shown in the figure 10.1. These data show that an 80 db SPL in a factory will effectively prevent conversation. Telephone conversations are similarly affected with a 65 db background making communication difficult and 80 db making it impossible.

Table 10.1. Sound levels for speech masking

Distance(M)	**Speech interference level (db)**	
	Normal	**Shouting**
1.	60	78
2.	54	72
4.	44	66

Example 10.1. A sound has an intensity of 5×10^{-8} w/m^2. What is sound level in decibel?

Solution:

$$db = 10 \log_{10} \left(\frac{I}{I_0}\right)$$

$$= 10 \log_{10} \left(\frac{5\times10^{-8}}{10^{-12}}\right)$$

$$= 10 \log_{10} 5 \times 10^4$$

$$= 40 \times 0.698$$

$$= 27.9 \text{ db.}$$

Example 10.2. If the intensity of sound is doubled, by how many decibels the sound level increased?

Solution:

$$db = 10 \log_{10} \left(\frac{I}{I_0}\right) \quad \text{... (1)}$$

If intensity of sound is doubled then

$$db = 10 \log_{10} \left(\frac{2I}{I_0}\right) \quad \text{... (2)}$$

Eqn. (2) – (1) $= 10 \log_{10}\left(\frac{2I}{I_0}\right) - 10 \log_{10}\left(\frac{I}{I_0}\right)$

$$= 10 \log_{10}\left(\frac{2I}{I_0}\right) = 10 \log_{10} 2 = 3 \text{ db.}$$

Example 10.3. What will be sound pressure level in decibel if rms sound pressure level is 20 N/m²?

Solution: $\text{db} = 20 \log_{10}\left(\frac{p}{p_0}\right) = 20 \log_{10}\left(\frac{20}{2\times 10^{-5}}\right)$

$$= 20 \log_{10} (10^6) = 20 \times 6 = 120 \text{ db.}$$

10.4. SOURCES OF NOISE

The sources of noise are more in urban and industrial areas, than in rural areas. The sources may be stationary or mobile. Some of the source are:

Industrial Source

Progress in technology (Industrialisation) has resulted in creating noise pollution. Textile mills, printing presses, engineering establishments and metal works etc. contribute heavily towards noise pollution.

In industrial cities like Kolkata, Ludhina, Kanpur etc. often the industrial zones are not separated from the residential zones of the city especially in the case of small scale industries. These operate from shops located on the ground floors of the residential area and cause annoyance, discomfort and irritation to the residents exposed to the noise that is inevitably produced.

The situation is much better in modern planned cities like Chandigarh where the industrial area is kept away from the residential areas.

Transport Vehicles

Increasing traffic has given rise to traffic jams in congested areas where the repeated hooting of horns by impatient drivers piere the ears of road users. Noise from aeroplanes constitute an increasingly serious problem in big cities like Delhi and Mumbai.

Household

The household is an industry in itself and is a source of many indoor noises such as the banging of doors, noise of playing children, crying of infants, moving of furniture, loud conversation of the inhabitants etc. Besides there are the entertainment equipment in the house, namely the radio, record players and television sets. Domestic gadgets like the mixer-grinders, pressure cookers, desert coolers, air-conditioners, exhaust fans, vacuum cleaner, sewing and washing machines are all indoor sources of noise pollution.

Public Address System

In India, people need only the slightest of an excuse for using loud-speakers. The reason may be a religious function, birth, death, marriage, elections, demonstrations or just commercial advertising. Public address systems therefore, contribute in its own way towards noise pollution.

Agricultural Machines

Tractors, thrashers, harvesters, tubewells, powered fillers etc. have all made agriculture highly mechanical but at the same time highly noisy. Noise level of 90 db to 98 db due to running of farm machines have been recorded in the state of Punjab.

Defence Equipments

A lot of noise pollution is added to the atmosphere by artillery tanks, launching of rockets, explosions, exercising of military, air planes and shooting practices.

Miscellaneous Sources

The automobile repair shops, construction works, blasting, bulldozing, stone crushing etc. are other sources of noise pollution.

10.5. EFFECTS OF NOISE

All sounds that are distracting, annoying, or harmful to everyday activities such as work, rest, entertainment, study and so forth are regarded as noises. Infact, any sound judged by the recipient as undesirable can be considered as a noise. Noise is generally harmful and a serious health hazard. It has far reaching consequences and have many physical, physiological as well as psychological effects on human being.

Whether or not a sound is undesired by a person will depend on a number of factors, *e.g.*

(*a*) Loudness.
(*b*) Frequency.
(*c*) Continuity.
(*d*) Variation with time.
(*e*) Time of occurrence.
(*f*) Information content.
(*g*) Origin of the sound.
(*h*) Recipient's state of mind and temperament.
(*i*) Background noise level.

In general, the effects of noises are:

(*a*) Physical effects

The physical manifestation of noise pollution is the effect on hearing ability. Repeated exposure to noise may result in temporary or permanent shifting of the hearing threshold of a person depending upon the level and duration of the exposure. The immediate and acute level of noise pollution is impairment of hearing (total deafness).

Human ears have sensory cells for hearing. If these cells are subjected to repeated sounds of high intensity before they have an opportunity to recover fully, they can become permanently damaged leading to impairment of hearing. Besides the sensory cells, the delicate tympanic membrane or the ear drum can also be permanently damaged by a sudden loud noise such as explosion.

(*b*) Physiological Effects

The physiological manifestation of noise pollution are several as mentioned below:

(*a*) Headache by dilating blood vessels of the brain.
(*b*) Increase in the rate of the heart beat.
(*c*) Narrowing of arteries
(*d*) Fluctuations in the arterial blood pressure by increasing the level of cholesterol in the blood.
(*e*) Decrease in heart output
(*f*) Pain in the heart.
(*g*) Digestive spasms through anxiety and dilation of the pupil of the eye, thereby causing an eye-strain.
(*h*) Impairment of night vision.
(*i*) Decrease in the rate of colour perception.
(*j*) Lowering of concentration and effect on memory.
(*k*) Muscular strain and nervous breakdown.

(*c*) **Psychological Effect**

The psychological manifestation of noise pollution are:

- Depression and fatigue which reduces the efficiency of a person.
- Insomnia as a result of lack of undisturbed and refreshing sleep.
- Straining of senses and annoyance as a result of slow but persistent noise from motorcycles, alarm clocks, call bells, telephone rings etc.

(*d*) Other Health Effects

People living or working in noisy environments develop the habit of shouting; some are stressed and become irritable. There is evidence that noise can lead to psychiatric disorders. When human beings are exposed to high noise levels, the blood vessels constrict and muscles tense.

(*e*) Speech Interference and Masking

One cannot effectively use speech communication in an environment in which the noise level is too high. Sometimes, the masking of warning shouts by background noise is responsible for industrial accidents.

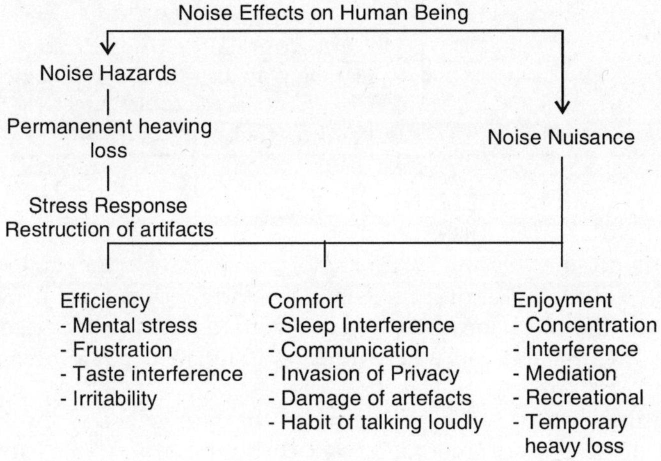

Fig. 10.2.

10.6. HUMAN AUDITORY SYSTEM

Sound pressure waves caused by vibration set the eardrum (tympanic membrane) in motion. This activates the three bones in the middle ear. The hammer, anvil and tirrup physically amplifying the motion received from the eardrum and transmit it to the inner ear. This fluid-filled cavity contains the cochea, a snail like structure in which the physical motion is transmitted to tiny hair cells. These hair cells deflect, much like sea weed swaying in the current, and certain cells are responsive only to certain frequencies. The mechanical motion of hair cells is transformed to bioelectrical signals and transmitted to the brain by the auditory nerves. Acute damage can occur to the ear drum by sudden loud noise. More serious is the chronic damage to the tiny hair cells in the inner ear. Prolonged exposure to noise of certain frequency pattern can cause other temporary hearing loss, which disappears in a few hours or permanent loss. The former is called temporary threshold shift and latter is known as permanent threshold shift. If summarised, the noise will have the following adverse effects:

1. It can upset the sense of balance.
2. It can reduce the efficiency on certain tasks.
3. It causes the pupil to constrict.
4. It can increase the blood pressure

Level (in db)	Effects
upto 23......	No disturbances
30 – 60........	Stress, tension, psychological (illness heart attack) effects especially at upper range
60 – 90........	Damages to health and topological (ear diseases)) effect.
Above 120.......	Painful effects in long run

10.7. NOISE MEASUREMENT

Need for Noise Measurement

Measuring noise levels and worker's noise exposures is the most important part of a workplace hearing conservation and noise control programme. It helps identify work locations where there are noise problems, employed who may be affected, and where additional noise measurements need to be made.

Identification of noise problems in the workplace is the first step towards noise measurement. A walk through survey helps in making this decision. The indicators of potentially hazardous noise level includes:

- noise is louder than busy city traffic.
- people have to raise their voice to talk to someone at one metre (3 feet) away.
- at the end of work shift people have to increase the volume of their radio or TV to a level too loud for others, and
- after working for a few years at the workplace, employees find it difficult to communicate in a crowd or party situation where there are other sounds or many voices.

Before taking field measurements, it is important to determine the type of information required. The person making the measurement must understand:

- the purpose of measurement compliance with noise regulations, having loss prevention, noise control, community annoyance etc.
- the sources of noise and times when the sources are operating.

Noise Measurement Instruments

The most common instruments used for measuring noise are the sound level meter (SLM), the integrating sound level meter (ISLM), and the noise dosimeter. It is important that you understand the calibration, operation and reading the instrument you use. The user's manual provided by the instrument manufacturer provides most of this information. Table 10.2 provides some instrument selection guidelines.

Table 10.2. Guidelines for Instrument Selection

Type of Measurement	Appropriate Instruments (in order of Preference)	Result	Comments
Personal noise exposure	1) Dosimeter	Dose or equivalent sound level	Most accurate for personal noise exposures
	2) ISLM*	Equivalent sound level	If the worker is mobile, it may be difficult to determine a personal exposure, unless work can be easily divided into defined activities.
	3) SLM**	dB(A)	If noise levels vary considerably, it is difficult to determine average exposure. Only useful when work can be easily divided into defined activities and noise levels are relatively stable all the time.
Noise levels generated by a particular source	1) SKM**	dB(A)	Measurement should be taken 1 to 3 metres from source (not directly at the source).
	2) ISLM**	Equivalent sound level dB(A)	Particularly useful if noise is highly variable; it can measure equivalent sound level over a short period of time (1 minute).
Noise survey	1) SLM	dB(A)	To produce noise map of an area; take measurements on a grid pattern.
	2) ISLM	Equivalent sound level dB(A)	For highly variable noise.
Impulse noise	1) Impulse SLM	Peak pressure dB(A)	To measure the peak of each impulse.

* SLM Stands for Sound Level Meter.

** ISLM stands for integrating Sound Level Meter.

1. Sound Level Meter (SLM)

The SLM consists of a microphone, electronic circuits and a readout display. The microphone detects the small air pressure variations associated with sound and changes them into electrical signals. These signals are then processed by the electronic circuitry of the instrument. The readout displays the sound level in decibels. The SLM takes the sound pressure level at one instant in a particular location.

To take measurement, the SLM is held at arm's length at the ear height for those exposed to the noise. With most SLMs it does not matter exactly how the microphone is pointed at the noise source. The instrument's instruction manual explains how to hold the microphone. The SLM must be calibrated before and after each use. The manual also gives the calibration procedure.

With most SLMs, the readings can be taken on either SLOW or FAST response. The response rate is the time period over which the instrument averages the sound level before displaying it on the readout. Workplace noise level measurements should be taken on SLOW response.

A Type 2 SLM is sufficiently accurate for industrial field evaluations. The more accurate and much more expensive Type 1 SLMs are primarily used in engineering, laboratory and research work. Any SLM that is less accurate than a Type 2 should not be used for workplace noise measurement.

An A-weighting filter is generally built into all SLMs and can be switched ON or OFF. Some Type 2 SLMs provide measurements only in dB(A), meaning that the A-weighting filter is ON permanently.

A standard SLM takes only instantaneous noise measurements. This is sufficient in workplaces with continuous noise levels. But in workplaces with impulse, intermittent or variable noise levels, the SLM makes it difficult to determine a person's average exposure to noise over a work shift. One solution in such workplaces is a noise dosimeter.

2. Integrating Sound Level Meter (ISLM)

The integrating sound level meter (ISLM) is similar to the dosimeter. It determines equivalent sound levels over a measurement period. The major difference is that an ISLM does not provide personal exposures because it is hand-held like the SLM, and not worn.

The ISLM determines equivalent sound levels at a particular location. It yields a single reading of a given noise, even if the actual sound level of the noise changes continually. It uses a preprogrammed exchange rate, with a time constant that is equivalent to the SLOW setting on the SLM.

3. Noise Dosimeter

A noise dosimeter is a small, light device that clips to a person's belt with a small microphone that fastens to the person's collar, close to an ear. The dosimeter stores the noise level information and carries out an averaging process. It is useful in industry where noise usually varies in duration and intensity, and where the person changes locations.

A noise dosimeter requires the following settings:

(*a*) **Criterion Level:** exposure limit for 8 hours per day five days per week. Criterion level is 90 dB(A) for many jurisdictions, 85 dB(A) some and 87dB(A) for Canadian federal jurisdictions.

(*b*) **Exchange rate:** 3 dB or 5 dB as specified in the noise regulation.

(c) **Threshold**: noise level limit below which the dosimeter does not accumulate noise dose data.

Wearing the dosimeter over a complete work shift gives the average noise exposure or noise dose for that person. This is usually expressed as a percentage of the maximum permitted exposure. If a person has received a noise dose of 100% over a work shift, this means that the average noise exposure is at the maximum permitted. For example, with a criterion level of 90 dB(A) and an exchange rate of 3 dB(A), an eight-hour exposure to 90 dB(A) a 100% dose. A four-hour exposure to 93 dB(A) is also a 100% dose, whereas an eight-hour exposure to 93 dB(A) is a noise dose of 200%.

Usually the manufacturer electronically adjusts dosimeters to the criterion level and exchange rate in use. You may have to adjust them to suit the exposure guidelines/standards in force in your jurisdiction. Osimeters also give an equivalent sound or noise level. This is the average exposure level for noise over the time dosimeter was on. It has the same total sound energy as the actual, variable sound levels to which a person is exposed over some time period. Scientific evidence suggests that hearing loss is affected by the total noise energy exposure. If a person is exposed over an eight-hour work shift to varying noise levels, it is possible to calculate an equivalent sound level which would equal the same total sound energy exposure. This would have the same effect on the person's hearing as the variable exposure actually received (Figure 10.3) cannot be analyzed-using this method. This type of spectrum will usually give sound-level readings that do not fit on the chairs. Similarly, the dotted portions of the curves are regions of poor accuracy of analysis.

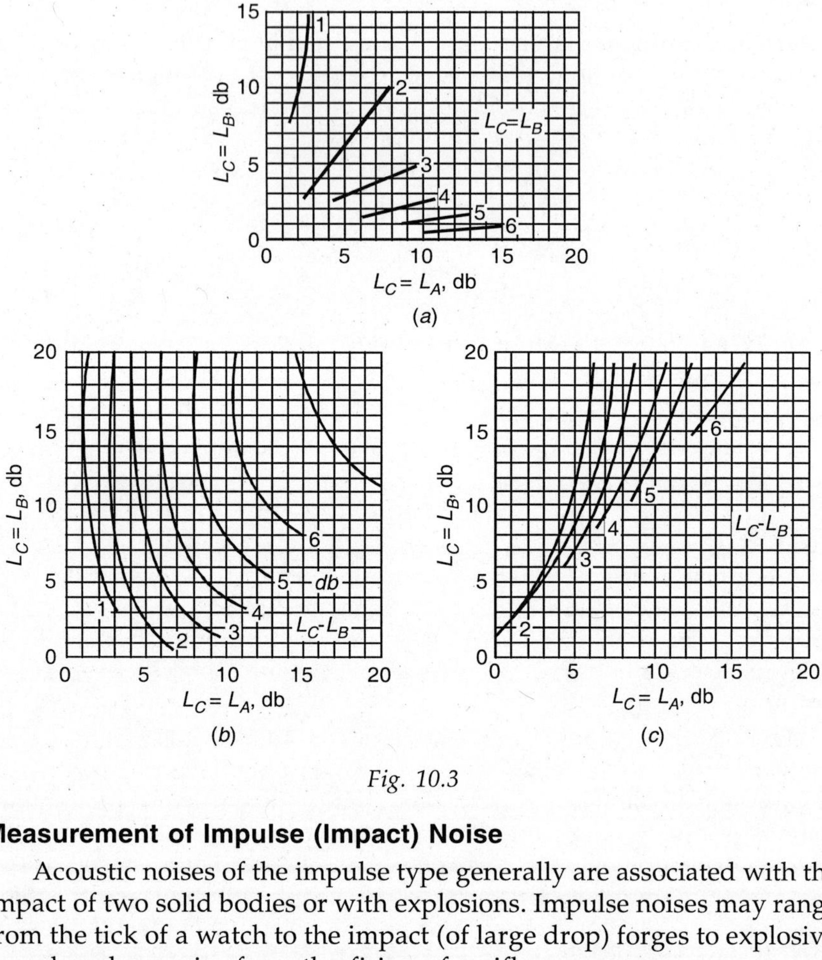

Fig. 10.3

Measurement of Impulse (Impact) Noise

Acoustic noises of the impulse type generally are associated with the impact of two solid bodies or with explosions. Impulse noises may range from the tick of a watch to the impact (of large drop) forges to explosive sound such as arise from the firing of a rifle or cannon.

Impulse noises are distinguished from continuous noises in two ways. First, the duration of the impulse noise is very short compared with the duration of the impulse noise is very short compared with the duration between occurrences of the impulse. Second, the peak value of the impulse noise is very much greater than the average or RMS value of the noise. Thus we say generally that a riveting gun is an impulsive noise source even though it is periodic.

The description of a single impulse noise is quite different from our description of continuous noises. In principle, an impulse noise is completely described by its time history, *i.e.*, a record of the instantaneous amplitude of the signal as a function of time, or by an energy-density spectrum.

The energy-density spectrum and a quantity related to the temporal behaviour of the instantaneous amplitude of a signal form a Fourier transform pair and are uniquely related to one another. These parameters are relatively difficult to obtain and to work with. In order to assess the effects of impact noises on hearing acuity, more simplified parameters are generally used.

If a simplifying assumption is made concerning the time history of impulse noises, two parameters may be used to provide a description of the physical characteristics of the noise. Let us assume that the rise time of the impulse noise is very rapid compared with typical time constants inherent in the hearing mechanism. Various experiments relative to the growth of loudness level, for example, indicate that the delay action in the middle car muscles is at least 10 msec. If the rise to the peak is much less than 10 msec, the rise time can be considered as infinitely fast. This assumption seems reasonable for most impact noises.

With this assumption a reasonable and useful description of the total energy in impact noise may be obtained from (1) the peak value of the noise and (2) some measure of its duration such as decay time. The decay time can be defined in the same way as is done in electrical circuits, namely, the time required for the envelope of the (exponentially) decaying wave to drop 10.7 db in level form its initial value. The sum of the peak instantaneous sound pressure (expressed in decibels) and ten times the logarithm of the decay time is proportional to the total energy in the impact noise, Such a sum will rank order impact noises on the basis of the total energy but does not give an absolute value of the total energy.

As an alternative, one may use an average value of sound pressure over some specified period of time to obtain another measure of the relative energy in impact 110/sec.

Some Measurement Procedures: As we have indicated above, there are several parameters which can be used to describe the physical characteristics of impulse noise. The physiological aspects of the relations of impulse-noises characteristics to loss of hearing acuity are, as yet, not defined well enough to form a basis of judgment concerning which parameters to use. The selection of a set of parameters to describe an impulse noise will therefore depend upon the instrumentation available to the engineer.

The measurement of impact noises is greatly facilitated by the use of a magnetic tape recorder. Use of a tape recorder allows a single measurement in the field to be analyzed in essentially any manner which may be desired at a later time. An octave band analysis of an impact noise in the field will require at least eight operations of the impact source. In practice, one might require ten or twelve operations to obtain one reading in each octave band. Analysis in the field without a tape-recorder will require many more operations of the source if the levels of the source vary significantly from operation to operation. It has been found that noise arising form the firing of rifles, machine guns, and cannons is very stable acoustically. However, stability may not be an attribute of other impact noises such as drop forges, punch presses, etc.

If a magnetic tape recorder is used, the overall peak value should be monitored. First, one obtains an approximate estimation of the stability of the sources. This observation is useful for estimating how many tape recordings will be required to obtain sufficiently stable average values of the characteristics of the noise. If, for example, the peak values for four consecutive operations are within 1 db of one another, there is little point in obtaining more than four recordings. If, on the other hand, the spread for four consective times is 6 db, then 10 or more operations of the impact source may be required to obtain a stable average value. The number of operations recorded will depend upon tolerable "error" in the final results.

Second, by monitoring one can prevent overload of the tape-recorder by the noise signal. The VU meter on the tape recorded does not respond rapidly enough to indicate a peak overload. The recorder can be instantaneously overloaded and the peak can be "clipped", even though the VU meter indicates a very low reading. For most industrial noise problems measurements should be made near the position of a machine operator's head, without the operator at his position. It may also be necessary to make measurements at other positions throughout the manufacturing space to determine the variation of noise intensity in space. Estimates of the variation in intensity should not be made on the basis of the usual room equations. The direct noise intensity cannot be accurately estimated form those equations because the propagation of sound for high peak levels is not a simple linear phenomenon. One finds that outdoor the peak levels drop off a rate of about 9 to 12 db per doubling of distance. The uncertainty in the propagation losses requires that the direct and reverberant fields be determined experimentally.

Magnetic tape recorders may be susceptible to microphonics induced by the high peak levels associated with typical impact noises. Whenever possible the tape recorder should be operated at a point that is distant from the noise source. If this not possible, the microscope should be removed from the record input, and the record attenuator should be set at the approximate level which will be used during the actual measurements. A recording should be made of the microphones output of the presence of the impact noise to determine if it is possible to obtain useful data during these conditions.

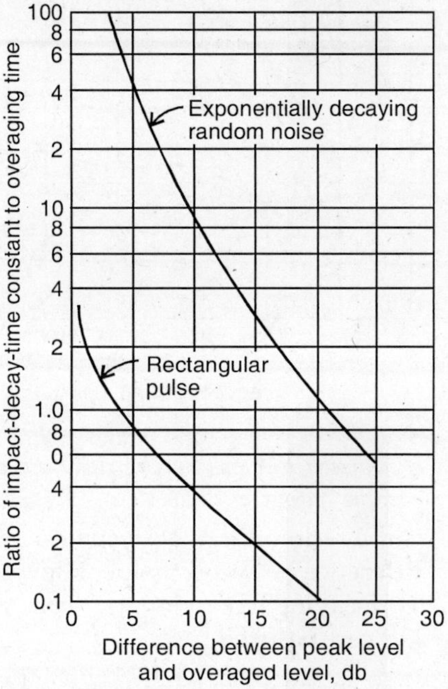

Fig. 10.4. Chart showing the relations between (1) the difference between the peak level (db) and the averaged level (db) and (2) the ratio of the decay time constant of the impact to the averaging time of the circuit. The lower curve is for a rectangular pulse, and the upper one is for usual impact noises.

Analysis of Impact Noise. The time history of an impact noise can be obtained from high-speed oscillographs or by photographing the trace on an appropriately calibrated oscilloscope. The energy-density spectrum can be obtained by appropriate filtering and integration. The analysis of an impact noise by use of an impact- noise meter, such as the General Radio type 1556A, is perhaps more interesting since the meter is relatively compact and can be used in the field or in conjunction with a tape-recorder.

The impact meter is a flexible device, and a significant amount of information can be obtained by proper use of it. The impact meter can be used to obtain (1) the peak level of the sound pressure and (2) a time-averaged level of the sound pressure. By measuring both the peak level and the averaged level, an estimate can be obtained of the duration (decay time constant) of the impact noise.

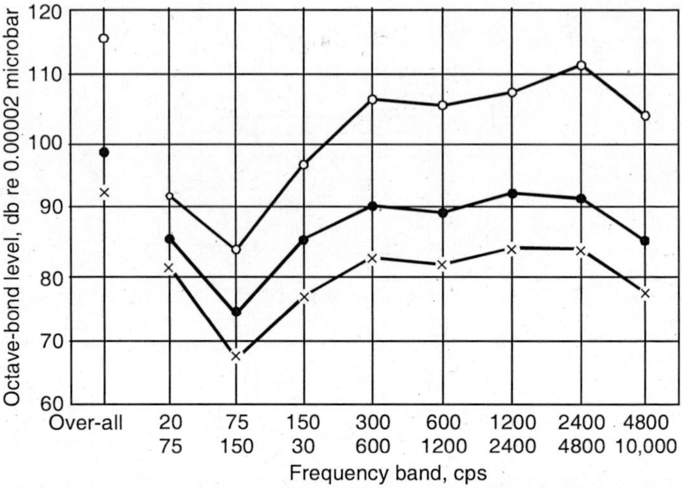

Fig. 10.5. The results of an octave band analysis of the noise from a single impact of a punch press as measured by the impact-noise analyzer on the output of the octave band analyzer.

Figure 10.5 shows the ratio of impact decay time to averaging time, as a function of the difference between the peak level and the averaged level, for exponentially decaying noise signals. A particular example will show how this relation can be used.

Measurements were made of a small punch press stamping out blanks. The measurements indicated a peak level of 115 db and a time-averaged level of 98 db when the "time constant" switch was set to 0.01 sec. The difference in level is 17 dB. From the upper curve of Fig. 10.4, it is found that this difference corresponds to a time ratio of 2. The equivalent impact decay time is then twice the averaging time or 0.02 sec. The noise from the punch press was also analyzed in octave bands of frequency. Both the peak value and the time-averaged value were obtained for each octave band.

● Measurement of Nonsteady Noise

A noise with a level that fluctuates over a range greater than 10 db during observation should be classified as nonsteady, either impulsive or fluctuating. An impulsive noise having a pulse repetition rate of greater than 10 pulses per second may usually be classified as a quasisteady noise, and the procedures outlined in the section on measurements of steady noise may be used. If the repetition rate is less than 10 pulses per second, however, the noise should be analyzed by the techniques described below.

• Fluctuating or Intermittent Noises

Cyclically or continuously fluctuating-noise or intermittent often must be measured with a sound level meter. The source of an intermittent noise is usually "on" (operating) for a period of time longer than the integration time of the ear (about 200 milliseconds). While "on", the noise may be constant in level or may fluctuate in level. If the noise is steady in level during the "on" time, one should record the central tendency of the "slow" meter setting during both the "on" and "off" times. For intermittent noises that fluctuate continuously during the "on" time, one should record the maximum and minimum levels rather than attempt to record central tendency. If the noise is cyclic in level, fluctuating between two or more well-defined levels, one may either record the central tendency at each level or tape-record the signal for analysis in the laboratory. If the noise fluctuates continuously in level, varying between an undefined maximum and minimum (such as heavy traffic on nearby thoroughfares), a statistical-distribution analyzer should be used.

• Fluctuating Noise (Less Than 10 dB Range)

When measuring noise that fluctuates over a range of less than about 3 dB it is accepted practice to observe the meter for approximately 10 sec and record a noise level representing the central tendency (average) of the need movement. If the fluctuations are greater than 3 dB but less than 10 dB an "average" according to Fig. 10.6 is recommended. For a more accurate statistical analysis, observe the meter deflection for about 5 sec and record both the "average" and the range in decibels of the meter fluctuations. Additional observations of the "average" should then be recorded until the number of observations is somewhat greater than the range in decibels of the meter fluctuations. The arithmetic mean of the "average" readings should be calculated and used as the noise level for that particular position. A more detailed statistical analysis can be performed by playing back a calibrated tape-recorded sample of the noise through appropriate filters either to a statistical-distribution or to a power-spectral-density analyzer.

• Ambient or Background Noise Measurements

When measuring ambient noise (the total noise associated with a given environment and usually comprising sounds from many sources both near and far) or background noise (the noise without a particular source in operation), the "slow" meter setting of the sound-level meter is generally used in conjunction with the procedures outlined above.

When measuring ambient noise levels for determining heating damage hazard, one of the microphone positions should be as close as possible to the normal location of the cars of the listener for whose benefit the measurement is being made, *e.g.*, a factory worker. The location will vary depending upon the activity of the listener.

DATE__________ SHEET _____ OF ______ PROJECT # __________ ENGINEER __________

CLIENT			
PROJECT			
OBSERVERS			
INSTRUMENTATION	SLM :	TYPE :	SER #
	TRANSDUCER :	TYPE :	SER #
	FILTER ANALYZER :	TYPE :	SER #
	CALIBRATOR	TYPE :	SER #
	OTHER :	TYPE :	SER #

PRIMARY NOISE SOURCE : __________
MANUFACTUERS NAME AND MACHINE DESIGNATION__________
OWNER'S DESIGNATION : __________
OPERATING CONDITIONS __________

SECONDARY NOISE SOURCE : __________
MANUFACTUERS NAME AND MACHINE DESIGNATION__________
OWNER'S DESIGNATION : __________
OPERATING CONDITIONS __________

TIME	CALIBRATION dB	°F dB	°F WB	% RH	min	Mg	W/No mph	WIMO Direction

TEST NO.	TIME HRS.	POSITION	CONDITIONS	SOUND-PRESSURE LEVEL dB RB 20 $\mu N/m^2/ms$										
				A WEIGHTED LEVEL	OVER-ALL LEVEL	OCTAVE BAND CENTER FREQUENCY, M								
						31.5	63	125	250	500	1k	2k	4k	6k

COMMENTS__________

Fig. 10.6. Suffused from for recording sound-pressure-level measurements.

The number of measurements required will depend upon the spatial distribution of the sound field and the purpose of the measurements. If the ambient noise level within a community is being established, it might be appropriate to determine the statistical properties of the noise, both the variations with time and with location. The number of measurement positions would depend upon the size of the community and the distribution of the noise sources within the community.

● Measurement of Steady Noise

Factors to Remember. The measurement of a steady noise involves the following few important factors:

1. Choose the proper equipment.
2. If a suitable analyzer is not available, take readings on all three scales of the sound-level meter so as to obtain some information on the distribution of the sound intensity with frequency.
3. Decide whether to use the slow or fast scale. If time-permits, take readings with both. As a general rule, if when using the fast scale

the fluctuations cover a 6 db or greater range, read 3 db down form the frequency peaks. If the fluctuations cover less than a 6 db range, read the average of the range. By this rule, usually the fast and slow-scale readings on fluctuating noises will be alike.

4. Calibrate the instrument according to the manufacturer's instructions at frequent intervals during each half-day of measurements. Preferably, an acoustical calibrator should be used.
5. Be certain that your microphone is placed (successively) at the locations where you want the noise measured and that you write down or sketch the locations exactly. Use as many measurement positions as are necessary to establish the characteristics of the noise being measured.
6. Orient the microphone properly in relation to the source.
7. Make certain that microphonics or background noises are not influencing the readings.
8. Read the stabilized temperature of the microphone frequently if it is of the crystal type or if the temperatures are very low or high.
9. Make sketches, notes and Tabulations of the year where the measurements are being performed, including sound-effecting and sound-absorbing surfaces.

Averaging Readings in Space. Frequently it is necessary to determine the average sound intensity radiated by a source through a hemispherical, spherical, or plane surface. If the assumption is made that the sound-pressure level is equal to the intensity level (as defined in Chap. 3), then the average IL can be determined from the average SPL. The simplest way of doing, this is to compute an average sound-pressure level (SPL) in decibels, by taking the arithmetic average of the SPL's in measuring sections of equal areas. Of course, some error will be present if this is done, except for the case in which all SPL's are equal. The value of SPL obtained by arithmetic averaging of decibels is always too low. The maximum error is approximately – 2.5 db for the case where there is an even distribution of the data over a 10 db spread for between two and ten measuring stations. Thus, the following rule of thumb may be adopted; If the spread of measured values of SPL at the measuring stations is around 10 db and if the SPL's in decibels are averaged, add 1 db to the average to get the SPL. The resulting SPL usually will be off not more than = 1 db. If the spread of measured values is 5 db or less, add no correction. Here also, the error will be less than 1db.

Readings on All Three Networks. In general, when no analyzer is used, it is recommended that readings on noises be taken with *all three* weighting positions. This procedure avoids ambiguities found in switching from one scale to another as mentioned above, and, at the same time, the three readings provide some, indication of the frequency distribution of the noise. If the level is essentially the same on all three networks, the sound probably predominates in frequencies above 600 cps. If the level is greater by several decibels on the C network than on the *A* and *B* networks, much of the noise is probably below 600 cps.

A more complete statement of this approximate analysis is given by the charts of Fig. 10.3, which can be used to give an approximation of the sound distribution in three frequency bands. These charts should be used only as a guide for determining in a preliminary way what the spectrum might be, and they should not be regarded as obviating a complete octave-band analysis. There are occasions, however, when it is impractical to make more than this preliminary analysis, and then the charts of Fig. 10.3 may help in making a more satisfactory decision about a noise problem than can be done with only one reading of a noise meter.

Certain noises in which the energy is localized at one end or the other of the lower and middle bands of this approximate analysis.

10.8. NOISE CONTROL

Noise is not merely a nuisance but is a serious environmental problem and a health hazard. Like all other pollutions, noise pollution needs to be control by measures which will maintain the acceptable levels of noise pollution for human beings and building. The control of noise is essential and which can be achieved by the following means.

1. ***Reducing the sound produced:*** Noise producing source should be rectified so that intensity of noise reduced at the source itself. There are many occasions where simple measures can provide sizable reduction in noise level. Some representative cases are given below:

2. ***Interrupting the path of the sound:*** Some barrier is made or the path of sound is channelised so that desired location can be segregated from the effect of the noise.

While planing an industrial complex the more noisy zones can be clubbed together at a reasonable distance from the less noisy zones if possible. This is called **zoning.**

By proper **Landscaping** a good amount of respite can be obtained from noise. This method may include provision of green verges or green belts, planting of tall leafy trees etc. Such arboriculture will act as good noise baffles and reduce the problems.

Source of Noise	Control Measure
1. Bearing make noise due to wear.	1. Change of bearings.
2. Excessive load on bearing.	2. Change the size of bearing.
3. Vibration and rattling of machine covers.	3. Shake proof fitting, use of rubber lining, tightening with spring washers and screws.
4. Mechanical arms carrying metal parts make noise.	4. Providing rubber lining to the arms.
5. Leaks in compressed air lines, steam pipes etc.	5. Stopping the leaks.
6. Noise by cutting tools.	6. Change of speed/feed, going for a harder tool, resharpening more frequently, shiffening the cutter holders.
7. Exhaust of diesel engines.	7. Install silencers.
8. Vibration of heavy rotors.	8. Dynamic balancing of rotors and better design of foundations.

The noisy factory area can be segregated from the office building of the factory by providing a suitable buffer zones.

The areas having more or less equal noise level may be located together if possible. This needs proper planning and called **clubbing**.

3. ***Protecting the recipient:*** The application of personal protection by using earmuffed or cotton plug. The adverse effects of noise can be controlled. In situation where neither source modifications, nor path modification reduces the noise to acceptable levels, direct protection of the individual is necessary. In industrial situations, this includes rotating staff members from one room to another, issuing ear protection equipment. Such as ear muffs, ear plugs and throw away ear plug material (like cotton); issuing personal noise dosimeters maintaining regular audiograms to monitor employee's hearing capabilities.

Industrial Noise Control

Industrial noise control generally involves the replacement of noise-producing machinery or equipment with quiet alternatives. For example, the noise from an air fan can be reduced by increasing the number of blades or the pitch of the blades and decreasing the rotational speed (thus obtaining the same air flow).

The second method of decreasing industrial noise is to interrupt the path, for example, by covering a noisy motor with insulating material.

The third method of noise control, often used in industry, is to protect the recipient by distributing earmuffs to the employees. But the problem is thus not really solved, and often the workers refuse to wear earmuffs, considering them sissy. There seems to be something masculine about being able to "take it". Such misdirected masculinity will only end up in deafness.

The occupational Safety and Health Act applies industrial noise limits to all jobs involving significant federal funds. These limits are tabulated in Table 10.3.

Table 10.3. Maximum Permissible Noise Level

Sound Level (dB)	Maximum duration in a day (hrs)
90	8
92	6
95	4
100	2
105	1
110	1/2
115	1/4

Community Noise Control

The three major sources of community noise are aircraft, highway traffic and construction activities.

Construction noise must be controlled by local ordinances (unless federal funs are involved). Control usually involves muffing or air compressors, jack hammers, hand compactors, etc. Since mufflers cost money, contractors will not take it upon themselves to control noise, and outside pressure must be exerted.

Aircraft noise in the United States is the Province of the Federal Aviation, Administration which has installed a two-pronged attack on this problem. First, it has set limits on aircraft engine noise and will not allow aircraft exceeding these limits to use the airports. This has forced manufacturers to design for quiet as well as for thrust.

The second effort has been to divert flight paths away from populated areas and whenever necessary to have pilots use less than maximum power when the takeoff carries them over a noise-sensitive area. Often this approach is not enough to prevent significant noise induced damage or annoyance, and aircraft noise remains a real problem in urban areas.

Supersonic aircraft present a special problem. Not only their engines are noisy, but the sonic boom can create property damage and mental anguish. The magnitude of this problem will become known only when (and if) supersonic airlines begin regular service over land.

The third major source of community noise is from highway. The car or truck create noise by a number of means:

1. Exhaust noise
2. Tire noise
3. Engine intake noise
4. Gears and transmission
5. Aerodynamic (wind) noise

A modern passenger car is well muffed that its most important contribution, a moderate and high speed, in tire noise. Sports car and motorcycles, on the other hand, contribute exhaust intake and gear noise.

The worst offenders on the roads is the heavy truck. Truck noise is generally from all the above sources. In most cases, the total noise generated by vehicles can be correlated directly to the truck volume.

A number of alternatives are available for reducing traffic noise. First, the source can be controlled by making quieter vehicles, second, highway could be routed away from populated areas; and third noise could be baffled with walls or other types of barriers. Other methods include lowering speed limits, designing for nonstop operation, and reducing all highways to less than 8% grade.

Vegetation, surprisingly, makes a very poor noise screen. The most effective buffers have been to raise or lower the highway, or to build physical barriers the road and thus screen the noise. All of these have limitations. For example, noise will bounce off the walls and create little or no noise shadow. In addition, walls hinder highway ventilation, thus contributing to the buildup of dangerous air pollutants.

The design noise level for various uses, are shown in Table 10.4.

Table 10.4. Design Noise Levels

Land Category	Design Noise Level	Description of Land Use
A	60 dB Exterior	Activities requiring special qualities of serenity and quiet, such as amphitheatres.
B	70 dB Exterior	Resistance, motels, hospitals, schools, parks, libraries.
	55 dB Interior	Residences, motels, hospitals, schools, parks, libraries.
C	75 dB Exterior	Developed land not covered in categories A and B.
D	No limit	Undeveloped land

Noise in the Home

Private dwellings get noisier because of internally produced noise as well as external community noise. The list of gadgets in modern home are the noisemakers. Some examples of domestic noise are listed in Table 10.5.

Table 10.5. Some Domestic Noisemakers

Item	Sound Level dB(A)
Vacuum cleaner (3 M)	75
Inside quiet car (80 KMPH)	65
Inside sports car (80 KMPH)	80
Flushing toiler	85
Garbage disposal (1 M)	80
Window air-conditioner (3 M)	55
Ringing alarm clock (0.6 M)	80
Lawn mower (operator's position)	105
Rock band (3 M)	115
Snowmobile (driver's position)	120

Similar products of different brands often will vary significantly in noise levels. When shopping for an appliance, it is just as important therefore to ask the clerk "How noisy is it" as it is to ask him "How much does it cost?". And if he looks at you as if you had two heads, explains to him that he should know the dB(A) at the operator's position for all or his wares. He may actually bother to find out.

10.9. NOISE INSULATION

Just like noise is transmitted by vibration through structures, sound waves travel through air and reach, the human ear. Substantial reduction in noise level can be achieved by insulating the source from the outside environment. Noise insulation is the word given to erecting a barrier (made of noise insulating material) in the path of noise. For best results, it is necessary to insulate the source from all the sides although insulating two or three sides only gives reduction, of a lower degree. Closing from all sides, involves problems relating to heat dissipation, accessibility for operation and maintenance, service facilities like air, water and electric connections and comparatively high cost. The reduction in noise level is achieved by the material reflecting the sound waves and will depend upon the insulating properties of the material used, the thickness of insulation maps or other path way left open in the barrier and the noise transmitted through path ways other than air.

10.10. NOISE ABSORPTION

Only Three Thing Stop Noise are:

1. **Space:** The more space there is, the more noise reduction there is. As an exaggeration, if you are 10 miles away, you will not hear it, except under strange circumstances.
2. **Mass:** A six foot thick concrete wall works great !! If is heavy, it will work well. Remember, if it does no weigh much it will not stop much.
3. **Dampening:** This requires a little explanation. Like a wine glass, when you make it ring ... but you can keep it quiet by holding if. You have dampened the noise by retarding the vibrations. This can be achieved in several ways.

Lead is an excellent noise stopper, its softness makes it to dampen the noise. Making noise transfer through different layers of material with different densities also helps dampen noise. Improperly done it can make things worse, like string holding chimes help them ring by isolating the vibration instead of dampening it.

10.11. NOISE REDUCTION Vs. NOISE ABSORPTION

This is a very important principle to understand about soundproofing. There are two types of noise reduction and they are almost always confused with each other. All noise reduction and noise Absorption are not alike. Noise absorption is not noise reduction.

Noise Reduction

Noise reduction is achieved by stopping it, killing it, not letting the noise get through the wall or window. Stop the noise before it gets to you, or into the room you are in. For this you use mass and space. You can build heavy walls with large airspaces and you maximize dampening as best you can.

Noise Absorption

Noise absorption is achieved by changing the characteristic of the noise. Stop it from echoing. Stop the reverberation. Stop the noise reflections. All this involves handling the noise that is within the room. It seldom involves noise from outside the room. You have likely heard of a room as being a 'live' room or a 'dead' room. For this type of noise reduction or soundproofing use carpets, upholstered furniture, acoustical ceiling tile, soundboard or interior walls.

Noise absorbent material normally soft and porous one, prevent reflection of noise and also convert some of the noise energy into heat energy. Auditoriums and Cinema theatres have their walls coloured with noise absorbent material with this purpose in view.

A combination of absorption and insulation can give very good reduction in noise level.

10.12. METHODS OF NOISE ABSORPTION

Acoustical materials exhibit a wide range of sound-absorbing efficiency over the audible spectrum, depending upon their surface treatment, composition, method of mounting, and thickness.

Nearly all the more common acoustical materials are of a porous composition with a surface finish designed to permit entrance of the sound wave into the interior. Their porous composition consists of a labyrinth of capillary-like tunnels which are formed by the interstices of fibrous mineral, cellulose, or aerated plaster products. When the materials is exposed to the sound wave, the air molecules contained in that part of the sound wave near the surface of the material partially migrate into the material in a rapid to-and-fro motion. Here a part of the acoustic energy is converted into heat through the frictional drag or viscous resistance encountered by the air molecules against the walls of the capillary tunnels. In this manner, the absorption of an incident acoustic wave can be as high as 95 per cent. The fraction of an incident energy absorbed (α, *absorption coefficient*) in the process will vary with frequency for a given material of a specified thickness and type of mounting.

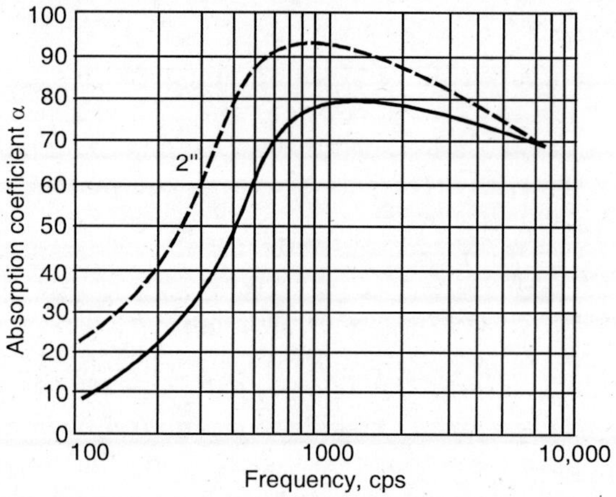

Fig. 10.7. Curves showing the increase in sound absorption resulting from a doubling in thickness of a typical homogeneous porous sound absorbing material.

Almost without exception, porous materials provide a greater percentage of sound absorption and the mid and high frequencies than at the lower frequencies. The lower curve of Fig. 10.7 illustrates a typical variation of absorption coefficient vs. frequency for a 1-in-thick "unfaced" low-density porous blanket with air-space backing. The upper curve of Fig. 10.7 shows that the same material having a 2-in thickness exhibits a significant increase in sound-absorbing efficiency, particularly at the lower and mid frequencies. While this is in fact a typical characteristics of porous acoustical materials, it should be pointed out that the increased sound-absorbing efficiency resulting from the increased thickness may not always be dramatic as that shown in Fig. 10.7.

Another material for sound absorption often used in architectural acoustics is the flexible impervious panel. When a panel is set into motion by an impinging sound wave, part of the energy contained in the wave is removed through internal viscous damping due to panel flexure.

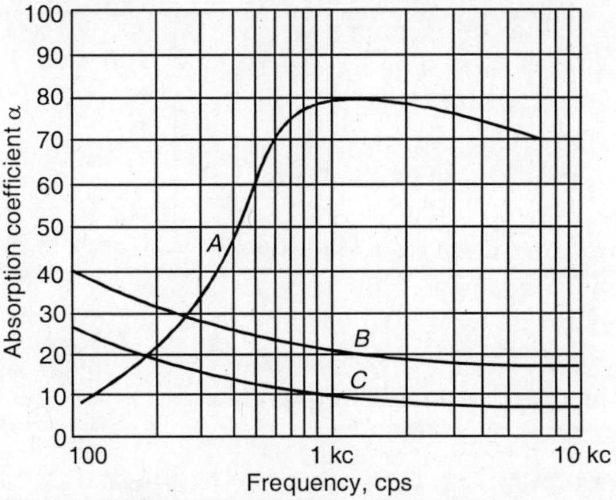

Fig. 10.8. Absorptiion coefficients for (A) ½-in thick porous acoustical aterial; (B) two sheets of ¾ thick plywood, separated by tufts of randomly glued felt causing ¼-in spacing between sheets (C) one sheet of $^{3}/_{8}$ -in-thick randomly braced plywood. The measurements of B and C were made in small broadcasting studios.

This is particularly true for frequencies in the region of the resonances of the panel. Panel materials such as plywood and composition hardboards of sizes normally used for interior finishes provide useful sound absorption, but we must note that they exhibit their maximum absorption in the lower frequency regions. On the other hand ½ to l-in-thick porous acoustical materials are somewhat less effective as these lower frequencies. Figure

10.8 shows the sound absorption as a function of frequency for two thicknesses of plywood panels in comparison to that of a typical 1-in-thick porous material. Maximum absorption is obtained with panels that are suitably damped and do not produce a ringing tone when struck. An example of an unsuitable panel material would be a thin undamped sheet of steel which is actually capable of re-emitting sound for a protracted period after the airborne source of acoustic excitation has been removed.

Still another method of achieving sound absorption is the *acoustic resonator of the Helmholtz type which consists of a vessel of any shape containing a volume of air*. This air volume is in direct communication with the air with in the room through an interconnecting tunnel which may be long or short and of any cross-sectional shape. An example of a resonator of this type might be a **gallon jug**. When a sound wave impinges on the aperture of the tunnel (neck of the jug), the air in the tunnel will be set into vibration which periodically expands and compresses the air within the vessel. The resulting amplified motion of the air particles in the tunnel due to phase cancellation between the air plough in the tunnel and the air volume in the vessels causes energy absorption due to frictional drag in and around the tunnel. This type of absorber can be designed to produce maximum absorption over a very narrow frequency range or even a wide frequency range as shown in Fig. 10.9.

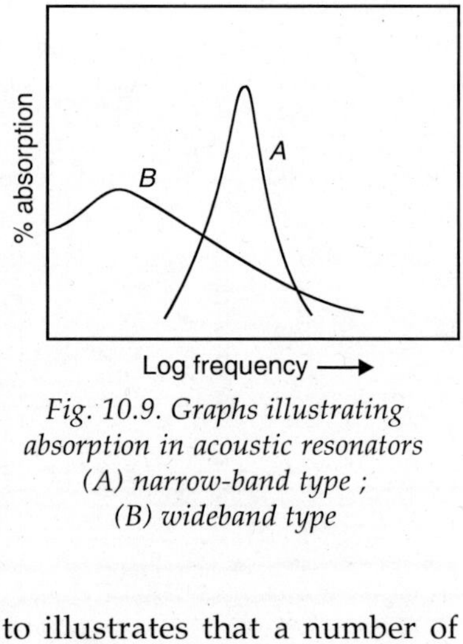

Fig. 10.9. Graphs illustrating absorption in acoustic resonators (A) narrow-band type ; (B) wideband type

The above brief descriptions serve to illustrates that a number of materials and even combinations of methods are available for absorbing sound over a broad or narrow frequency range and selectively at low, middle, and high frequencies.

Prefabricated Acoustical Materials

In recent years has been an ever-increasing demand for architectural acoustical materials. Competition among manufacturers has resulted in a significant increase in the number of products. The abundance of products which are now available in a wide variety of basic materials, thicknesses, tile sizes, (board sizes) and surface finishes makes the selection of an appropriate material somewhat difficult. Figure 10.4 shows a sampling of a number of surface finishes.

Fortunately, nearly all the prefabricated acoustical materials now available have been typed in accordance with there general appearance and composition by the Acoustical Materials Association (AMA). This association is composed of a number of nationwide organization engaged in the manufacture of architectural acoustical products. Particular emphasis has been directed by the association toward maintaining or improving the absorption characteristics of various materials while simultaneously offering a greater variety of decorative features. Table 10.4 offers a list of the materials as classified by the AMA.

As mentioned earlier, the thickness and method of mounting are of equal importance in determining the sound absorption coefficient of a material at various frequencies. The prefabricated products have been standarized essentially into ½ in. ¾ in.. and 1 in., thicknesses by the AMA. The methods of application or types of mountings used in conducting tests to determine sound-absorption coefficient.

10.13. SILENCER

A sound suppressor is designed to be most effective with a given firearm and cartridge. The main purpose of a sound suppressor is to reduce the muzzle blast, which is the most significant portion of the noise. Muzzle blast is caused by high pressure gases suddenly escaping from the end of the barrel into the atmosphere as the bullet exits. Reducing the pressure results in less sound generated. Pressure is reduced by increasing the volume for gas expansion, reducing the gas temperature (cooling) delaying gas exit through trapping and turbulence, and by a combination of these technologies.

Types of Suppressors/Silencers

There are two basic types of suppressors. The most common, and oldest, historically, is the muzzle suppressor which is attached to the muzzle of a firearm. The other type of suppressor is the integral suppressor, which is built around the barrel of a firearm. When using an integral type suppressor the barrel is modified for velocity control.

Silencers are more properly called sound moderators or suppressors and are widely used by hunters to reduce noise levels from the discharge of firearms, particularly sporting rifles and air weapons. A sound moderator is essentially no more than a series of baffles coupled to an expansion chamber, contained within a tubular attachment which screws on to the end of the firearm's barrel.

The noise of the discharge of most firearms is made up of two components. The first comes from the rapid expansion of propellant gases as they leave the muzzle. The second is the supersonic crack of the bullet. It is not possible to reduce the sound level of a supersonic bullet, but a sound moderator fitted to such a rifle will have some significant effect in reducing the noise signature because it controls the rate of expansion of the propelling gases.

For a sound moderator to be really effective, it must be used with ammunition whose projectiles travel at less than the speed of sound. In such cases, the noise of the discharge is greatly reduced and may not even be recognisable as a gun.

It is not possible to fit a sound moderator to a revolver because the gap between the barrel and the front of the cylinder means that about 5 per cent of the propellant gases escape, contributing to the overall noise of the discharge. Otherwise, sound moderators can be fitted to any type of firearm.

Working of Gun Silencer

It is amazing that anything is able to silence a gun, but gun silencers actually work on a very simple principle.

Imagine a balloon. If you pop a balloon with a pin, it will make a loud noise. But if you were to untie the end of the balloon and let the air out slowly, you could pop it making very little noise. That is the basic idea behind a gun silencer.

To fire a bullet from a gun, gunpowder is ignited behind the bullet. The gunpowder creates a high-pressure pulse of **hot gas**. The pressure of the gas forces the bullet down the barrel of the gun. When the bullet exits the end of the barrel, it is like uncorking a bottle. The pressure behind the bullet is immense, however — on the order of 3,000 pounds per square inch (psi) — so the POP that the gun makes as it is uncorked is extremely loud.

A silencer screws on to the end of the barrel and has a **huge volume** compared to the barrel (20 or 30 times greater). With the silencer in place, the pressurized gas behind the bullet has a big space to expand into. So the **pressure** of the hot gas falls significantly. When the bullet finally exits through the hole in the silencer, the pressure being uncorked is much, much lower — perhaps 60 psi. Therefore, the sound of the gun firing is much softer.

10.14. NOISE POLLUTION CONTROL LAWS

As the increasing ambient noise levels in public places from various sources, inter-alia, industrial activity, construction activity, generator sets, loud-speakers, public address systems, music systems, vehicular horns and other mechanical devices have deleterious effects on human health and the psychological well-being of the people, it is considered necessary to regulate and control noise producing and generating sources with the objective of maintaining the ambient air quality standards in respect of noise.

To have a legal standing, under the Environment (Protection) Act, 1986 (29 of 1986). The Noise Pollution (Regulation and Control) Rules, 2000 has been prepared and notified in the Gazette of India. Some of the important provisions contained in the said Rules are given below:

1. Ambient air quality standards in respect of noise for different areas/zones.

(*i*) The ambient air quality standards in respect of noise for different areas/zones shall be such as specified in the Schedule annexed to these rules.

(*ii*) The State Government may categories the areas into industrial, commerical, residential or silence areas/zones for the purpose of implementation of noise standards for different areas.

(*iii*) The State Government shall take measures for abatement of noise including noise emanating from vehicular movements and ensure that the existing noise levels do not exceed the ambient air quality standards specified under these rules.

(*iv*) All development authorities, local bodies and other concerned authorities while planning development activity or carrying out functions relating to town and country planning shall take into consideration all aspects of noise pollution as a parameter of quality of life to avoid noise menace and to achieve the objective of maintaining the ambient air quality standards in respect of noise.

(*v*) An area comprising not less than 100 metres around hospitals, educational institutions and courts should be declared as silence area/zone for the purpose of these rules.

2. Responsibility as to Enforcement of Noise Pollution Control Measures:

(*i*) The noise levels in any area/zone shall not exceed the ambient air quality standards in respect of noise as specified in the SCHEDULE.

(*ii*) The authority shall be responsible for the enforcement of noise pollution control measures and the due compliance of the ambient air quality standards in respect of noise.

SCHEDULE

Ambient Air Quality Standards in respect of Noise

Area Code	Category of Area/Zone	Limits in dB(A) Leq*	
		Day Time	Night Time
A	Industrial area	75	70
B	Commerical area	65	55
C	Residential area	55	45
D	Silence zone	50	40

Notes 1. Day time shall mean from 6.00 am to 10.00 pm

2. Night time shall mean form 10.00 pm to 6.00 am
3. Silence zone is defined as an area comprising not less than 100 metres around hospitals, educational institutions and courts. The silence zones are zones which are declared as such by the competent authority.
4. Mixed categories of areas may be declared as one of the four above mentioned categories by the competent authority.

* dB(A) Leqdenotes the time weighted average of the level of sound in decibels on scale A which is relatable to human hearing.

A "decibel" is a unit in which noise is measured.

"A", in dB(A) Leq denotes the frequency weighting in the measurement of noise corresponds to frequency response characteristics of the human ear.

Leq. It is an energy of the noise level over a specified period.

3. Restrictions on the Use of Loud Speakers/Public Address System

(*i*) A loud speaker or a public address system shall not be used except after obtaining written permission from the authority.

(*ii*) A loud speaker of a public address system shall not be used at night (between 10.00 pm to 6.00 am) except in closed premises for communication within, *e.g.*, auditorium, conference rooms, community halls and banquet halls.

4. Consequences of Any Violation in Silence Zone/Area

Whoever, in any place covered under the silence zone/area commits any of the following offence, he shall be liable for penalty under the provisions of the Act:

(*i*) whoever, plays any music or uses any sound amplifiers, (*ii*) Whoever, beats a drum or tom-tom or blows a horn either musical or pressure, or trumpet beats or sounds any instrument, or

(*iii*) whoever, exhibits any mimetic, musical or other performances of a nature to attract crowds.

IMPORTANT QUESTIONS

1. Define noise. Discuss continuous and impulse noise.
2. Does noise affect the man is working? Explain.
3. What is noise isolation? What are different noise evaluation procedures? Explain them in detail.
4. What are various noise absorption techniques?
5. What are sources of noise? Explain.
6. What are various measures used to control noise?
7. Write a short note on the working of silencers.
8. State the purpose of silencer.

CHAPTER 11

VIBRATIONS

11.1. INTRODUCTION

If you could watch a vibrating object in slow motion, you can see movements in different directions. How far, and how fast the object moves helps determine its vibrational characteristics. The terms used to describe this movement are frequency, amplitude and acceleration.

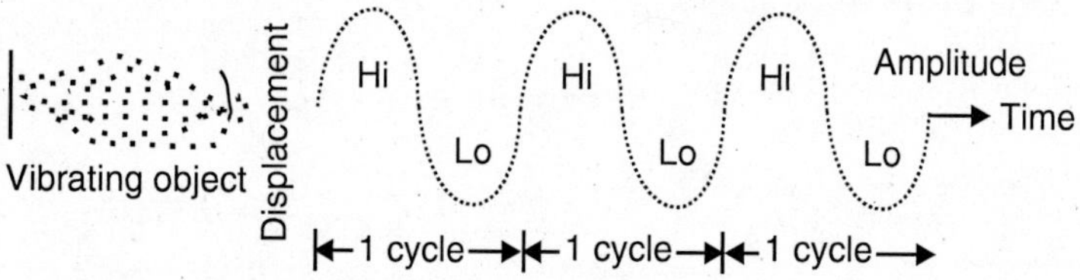

Fig. 11.1.

Frequency

A vibrating object moves back and forth from its normal stationary position. A complete cycle of vibration occurs when the object moves from one extreme position to the other extreme, and back again. The number of cycles that a vibrating object completes in one second is called frequency. The unit of frequency is hertz (Hz). One hertz equals one cycle per second.

Amplitude

A vibrating object moves to a certain maximum distance on either side of its stationary position. Amplitude is the distance from the stationary position to the extreme position on either side and is measured in metres (m). The intensity of vibration depends on amplitude.

Acceleration

The speed of a vibrating object varies from zero to a maximum during each cycle of vibration. It moves fastest as it passes through its stationary position to an extreme position. The vibrating object slows down as it approaches the extreme, where it stops and then moves in the opposite direction through the stationary position toward the other extreme. Speed is expressed in units of metres per second (m/s).

Acceleration is a measure of how quickly speed changes with time and therefore, acceleration is expressed in units of metres per second squared (m/s^2). The magnitude of acceleration changes from zero to a maximum during each cycle of vibration. It increases as vibrating object moves further from its normal stationary position.

Resonance

Every object tends to vibrate at one particular frequency that depends on the composition of the object, its size, structure, weight and shape. This frequency of natural vibration is called the resonant frequency. A vibrating machine transfers the maximum amount of energy to an object when the machine vibrates at the object's resonant frequency.

11.2. VIBRATION EXPOSURE

Vibration exposure is possible in many occupations where a worker comes in contact with vibrating machinery or equipment. When a worker operates hand-held equipment such as a chain saw or jackhammer, vibration affects hands and arms and is called **hand-arm vibration exposure**. When a worker sits or stands on a vibrating floor or seat, the exposure affects almost the entire body and is called**whole-body vibration exposure**. The risk of injury depends on the intensity and frequency of the vibration, the duration (years) of exposure and the part of the body which receives the vibration energy.

Hand-arm vibration causes damage to blood vessels and nerves in the fingers. The resulting condition is known as white finger disease, Raynaud's phenomenon or hand-arm vibration syndrome (HAVS). The affected fingers turn white, especially when exposed to cold. Vibration-induced white finger disease also causes a loss of grip force and reduced sensitivity to touch.

The effect of whole-body vibration is poorly understood. Studies of drivers of heavy vehicles have revealed an increased incidence of the disorders of bowel and the circulatory, musculoskeletal and neurological systems.

However, disorders of the nervous, circulatory and digestive systems are not specific to whole-body vibration exposure. These disorders are assumed to be caused by a combination of various working conditions and life style factors rather than by one physical factor alone.

11.3. EFFECT OF HAND-ARM VIBRATION

Vibration-induced white finger (VWF) is the most common condition among the operators of hand-held tools. Vibration can cause changes in tendons, muscles, bones and joints, and can affect the nervous. Collectively, these are known as Hand-Arm Vibration Syndrome (HAVS). The symptoms of vibration aggravated when the hands are exposed to cold.

Workers affected by HAVS commonly report:

- attacks of whitening (blanching) of one or more fingers when exposed to cold
- tingling and loss of sensation in the fingers
- loss of light touch
- pain and cold sensations between periodic white finger attacks
- loss of grip strength
- bone cysts in fingers and wrists

The development of HAVS is gradual and increases in severity over time. It may take a few months to several years for the symptoms of HAVS to become noticeable.

Symptoms of Hand-arm Vibration Syndrome (HAVS)

Hand-arm vibration exposure affects the blood flow (vascular effect) and causes loss of touch sensation (neurological effect) in fingers. One of the earliest methods used for identifying the severity of these symptoms was the Taylor-Pelmear classification method. Table 11.1 shows the Taylor-Pelmear classification of the clinical stages of vibration induced white finger. This classification was widely used in the past.

Table 11.1. Taylor-Pelmear classification of vibration-induced white finger by stage

Stage	Signs and Symptoms	Interference with Activities
0	None	None
OT	Intermittent tingling	None
ON	Intermittent numbness	None
OTN	Tingling and numbness	None
1	Blanching of one or more fingertips with or without tingling and numbness	None
2	Blanching of one or more fingers with numbness, usually during winter only	Slight interference with home and social activities interference with work
3.	Extensive blanching with frequent episodes during both summer and winter	Definite interference with work, home and social restricted hobbies
4.	Extensive blanching of most fingers; frequent episodes during summer and winter; finger ulceration	Occupation change required to avoid further vibration exposure

In 1986, a new classification, known as Stockholm classification was introduced. In this classification vascular (blood flow) changes and neural (feeling of touch, heat, cold, etc.) changes are considered separately [see Tables 11.2 and 11.3].

Table 11.2. The Stockholm workshop classification scale for cold-induced vascular (blood flow) symptoms in fingers with hand-arm vibration syndrome

Stage	Grade	Description
0	(None)	No attacks
1	Mild	Occasional attacks affecting only the tips of one or more fingers
2	Moderate	Occasional attacks affecting fingertips and middle of the finger and rarely also the finger
3	Severe	Frequency attacks affecting most fingers
4	Very Severe	Same symptoms as in stage 3 with degenerate skin changes in the fingertips.

Table 11.3 (*b*). The Stockholm workshop classification scale for sensorineural change in fingers due to hand-arm vibration syndrome

Stage	Symptoms
OSN	Exposed to vibration but no symptons
1SN	Intermittent numbness, with or without tingling
2SN	Intermittent or persistent numbness, reduced sensory perception
3SN	Intermittent or persistent numbness, reduced tactile discrimination and/ or manipulative dexterity

The severity of hand-arm vibration syndrome depends on several other factors, such as the character vibration exposure, work practice, personal history and habits. Table 11.4 summarizes these factors.

Table 11.4.
Factors that influence the effect of vibration on the hand

Physical Factors	Biodynamic Factors	Individual Factors
Acceleration of vibration	Grip forces - how hard the worker grasps the vibrating equipment	Operator's control of tool
Frequency of vibration	Surface area, location, and mass of parts of the hand in contact with the source of vibration.	Machine work rate
Duration of exposure each workday	Hardness of the material being contacted by the hand - held tools, for example metal in grinding and chipping.	Skill and productivity
Years of employment involving vibration exposure	Position of the hand and arm relative to the body.	Individual susceptibility to vibration
State of tool maintenance	Texture of handle-soft and compliant versus rigid material	Smoking and use of drugs. Exposure to other physical and agents.
Protective practices and equipment including gloves, boots, work-rest periods.	Medical history of injury to fingers and hands, particularly frostbite	Diseases or prior injury to the fingers hands.

Hand-arm vibration syndrome is also known as **Raynaud's phenomenon** of occupational origin. Vibration is just one cause of Raynaud's phenomenon. Other causes are connective tissue diseases, tissue injury, diseases of the blood vessels in the finger, exposure to vinyl chloride, and the use of certain drugs. The resulting reduced blood flow can produce white fingers in cold environments.

11.4. EFFECT OF WHOLE-BODY VIBRATION

Whole-body vibration can cause fatigue, insomnia, headache and "Shakiness" shortly after or during exposure. The symptoms are similar to those that many people experience after a long car or boat trip. After daily exposure over a number of years, whole-body vibration can affect the entire body and result in a number of health disorders. Sea, air or land vehicles cause motion sickness when the vibration exposure occurs in the 0.1 to 0.6 Hz frequency range. Studies of bus and truck drivers found that occupational exposure to whole-body vibration could have contributed to a number of circulatory bowel, respiratory, muscular and back disorders. The combined effects of the body posture, postural, fatigue, dietary habits and whole-body vibration are the possible causes for these disorders.

Studies show that whole-body vibration can increase heart rate, oxygen uptake and respiratory rate produce changes in blood and urine. East European researchers have noted that exposure to whole-body vibration can produce an overall ill feeling which they call "Vibration Sickness".

Many studies have reported decreased performance in workers exposed to whole body vibration.

Whole-Body Vibration energy enters the body through a seat or the floor, it affects the entire body or a number of organs in the body. Exposed of groups include operators of trucks, buses, tractors and those who work on vibrating floors. Table 11.5 lists examples of vibration exposure in various industries.

Table 11.5 Examples of Occupational Vibration Exposure

Industry vibration	**Type of vibration**	**Common source of**
Agriculture	Whole-body	Tractors
Boiler making	Hand-arm	Pneumatic tools
Construction	Whole-body	Heavy equipment vehicles
	Hand-arm	Pneumatic tools, jack hammers
Diamond cutting	Hand-arm	Vibrating hand tools
Forestry	Whole-body	Tractors
	Hand-arm	Chain saws
Foundries	Hand-arm	Vibrating cleavers
Furniture manufacture	Hand-arm	Pneumatic chisels
Iron and steel	Hand-arm	Vibrating hand tools
Lumber		Hand-arm Chain saws
Machine tools	Hand-arm	Vibrating hand tools
Mining	Whole-body	Vehicle operation
	Hand-arm	Rock drills
Rivetting	Hand-arm	Hand tools
Rubbers	Hand-arm	Pneumatic stripping tools
Sheet metal	Hand-arm	Stamping equipment
Shipyards	Hand-arm	Pneumatic hand tools
Shoe-making	Hand-arm	Pounding machine
Stone dressing	Hand-arm	Pneumatic hand tools
Textile	Hand-arm	Sewing machines, Looms
Transportation	Whole-body	Vehicle

11.5. LIMITS OF VIBRATION EXPOSURE

As in all occupational exposures, individual sensitivity to vibration varies from person to person. Three important factors to be considered for vibration exposure are:

- The threshold value.
- The exposure-response relationship.
- Latent period.

The threshold value of vibration is the level below which there is no risk of vibration syndrome. In other words, it is the maximum intensity of vibration to which workers can be exposed every workday for their entire full-time employment without developing numbness, paleness or chill of fingers. Workers will not develop vibration-related injuries or disease if their exposure to vibration is maintained at sufficiently low levels.

The number of affected people increases as the intensity and duration of vibration exposure increases. This type of exposure-response relationship indicates a possible link between health effects and the total amount of vibration energy entering the hands or body. Depending on the intensity of exposure the symptoms may appear months or years after the start of the exposure.

The latent period for VWF is the time from the first occupational exposure to hand-arm vibration until the onset of symptoms. The latent period depends on the intensity of exposure. The higher the intensity, the shorter the latent period. Table 11.6 shows typical latent periods reported for workers in some occupations.

Table 11.6
Average latent periods for vibration-induced Diseases in different occupations

Occupation	Stage of VWF	Latency (Years)
Foundry worker	Tingling	1.8
	Numbness	2.2
	Blanching	2.0
Shipyard worker	Tingling	11.1
	Numbness	12.0
	Blanching	16.8
Chain saw operator	Numbness	4
Grinder	Blanching	13.7

11.6. ACCEPTANCE OF VIBRATION SYNDROME

The acceptance of vibration syndrome as an industrial syndrome is hindered because:

- Not every physican is trained to diagnose vibration - induced white finger (VWF) as other vibration diseases.
- The cause of VWF cannot always be identified.
- There are no objective clinical tests to measure the impairment.
- The disease progresses for years before the symptoms become severe enough to affect a worker to do her or his work.

11.7. MEASUREMENT OF VIBRATION

A complete assessment of exposure to vibration requires the measurement of acceleration in well-defined directions, frequencies and duration of exposure. How hard a person grips a tool affects the amount of vibrational energy entering the hands; Thereof, hand-grip force is another important factor in the exposure assessment.

Most jurisdictions and agencies use acceleration as a measure of vibration exposure for the following reasons:

- Several types of instruments are available for measuring acceleration, the rate of change of velocity or speed or direction per unit time.
- Measuring acceleration can also give information about velocity and amplitude of vibration.
- The degree of harm is related to the magnitude of acceleration.

A typical vibration measurement system includes a device to sense the vibration (accelerometer), a tape recorder, a frequency analyser, a frequency-weighting network, and a display such as a meter, printer or recorder.

The accelerometer produces an electrical signal. The size of this signal is proportional to the acceleration applied to it.

The frequency analyzer determines the distribution of acceleration in different frequency bands. The frequency-weighting network measures the human sensitivity to vibration of different frequencies. The use of weighting networks gives a single number as a measure of vibration exposure and is expressed as the frequency-weighted vibration exposure in meters per second squared (m/s^2), units of acceleration. Vibration meters equipped with accelerometers can be used for instantaneous vibration measurement. Some types of sound level meters can measure vibration.

The frequency-weighting network for hand-arm vibration is given in the International Organization for Standardization (ISO) standard ISO 5349. Human hand is not equally sensitive to vibration energy at all frequencies. The sensitivity is the highest around 8-16 Hz (Hertz or cycles per second). Measuring equipment takes this fact into account by using a weighting network. The gain is assigned a value of 1 for vibration frequencies to which the hand-arm system has the highest sensitivity. The dashed lines in figure 11.2 represent the filter tolerance in the weighting network.

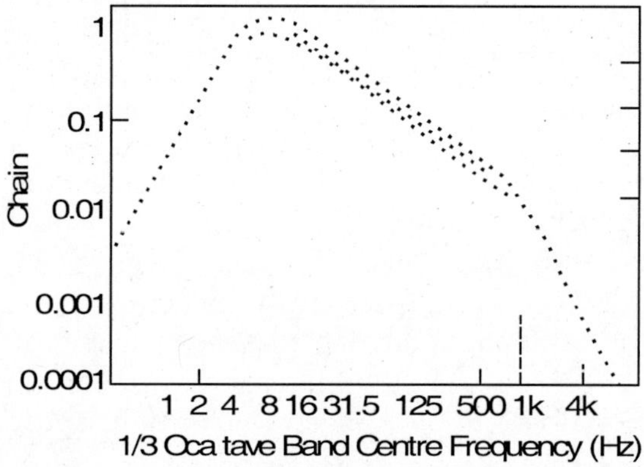

Fig. 11.2.

11.8. METHODS FOR CONTROLLING EXPOSURE TO VIBRATION

Protecting workers from the effects of vibration usually requires a combination of appropriate tool selection, the use of appropriate vibration-absorbing materials (in gloves, for example), good work practices, and education programmes.

(i) Anti-vibration Tools

Using anti-vibration chain saws reduces acceleration levels by a factor of about 10. These types of chain saws must be well maintained. Maintenance must include periodic replacement of shock absorbers. A few pneumatic tool companies manufacture anti-vibration tools such as anti-vibration pneumatic chipping hammers, pavement breakers and vibration-damped pneumatic riveting guns.

(ii) Anti-vibration Gloves

Conventional protective gloves (*e.g.*, cotton, leather), commonly used by workers do not reduce the vibration that is transferred to workers' hands when they are using vibrating tools or equipment. Anti-vibration gloves are made using a layer of viscoelastic material. Actual measurements have shown that such gloves have limited effectiveness in absorbing low-frequency vibration, the major contributor to vibration-related disorders. Therefore, they offer little protection against developing vibration-induced white finger syndrome. However, gloves do provide protection from typical industrial hazards (*e.g.*, cuts, abrasions) and from cold temperatures that, in turn, may reduce the initial sensation of white finger attacks.

(iii) Safe Work Practices

Along with using anti-vibration tools and gloves, workers can reduce the risk of hand-arm vibration syndrome (HAVS) by following work practices:

- Employ a minimum hand grip consistent with safe operation of the tool or process.
- Wear sufficient clothing, including gloves, to keep warm.
- Avoid continuous exposure by taking rest periods.
- Rest the tool on the work piece whenever practical.
- Refrain from using faulty tools.
- Maintain properly sharpened cutting tools.
- Consult a doctor at the first sign of vibration disease and ask about the possibility of changing to a job with less exposure.

(iv) Employee Education

Training programmes are an effective means of heightening the awareness of HAVS in the workplace. Training should include proper use and maintain vibrating tools to avoid unnecessary exposure to vibration. Vibrating machines and equipment often produce loud noise as well. Therefore, training and education in controlling vibration should also address concerns about noise control.

(v) Whole-Body Vibration

The following precautions help to reduce whole-body vibration exposure:

- Limit the time spent by workers on a vibrating surface.
- Mechanically isolate the vibrating source or surface to reduce exposure.
- Ensure that equipment is well maintained to avoid excessive vibration.
- Install vibration damping seats.

The vibration control design is an intricate engineering problem and must be set up by qualified professionals. Many factors specific to the individual work station govern the choice of the vibration isolation material and the machine mounting methods.

11.9. STANDARDS OR GUIDELINES FOR EXPOSURE TO HAND-ARM VIBRATION

The American Conference of Governmental Industrial Hygienists (ACGIH) has developed Threshold Limit Values (TLVs) for vibration exposure from hand-held tools. The exposure limits are given as frequency-weighted acceleration that represents a single number measure of the vibration exposure level. The frequency-weighting is based on a scheme recommended in the international standard ISO 5349. Vibration-measuring instruments have a frequency-weighting network as an option for vibration measurement. Table 11.7 lists acceleration levels and exposure durations to which, ACGIH has determined, most workers may be exposed repeatedly without severe damage to fingers. ACGIH advises that these guidelines be applied in conjunction with other protective measures including vibration control.

Table 11. 6

The ACGIH Threshold Limit Values (TLVs) for exposure of the hand to vibration in X, y, or z direction*

Total Daily Exposure Duration (hours)	Maximum values of frequency weighted acceleration (m/s^2) in any direction*
4 to less than 8 hours	4
2 to less than 4 hours	6
1 to less than 2 hours	8
less than 1 hour	12

* Directions of axes in the three-dimensional, system.

The International Organization for Standardization (ISO) has published a method for measuring vibration and interpreting the resulting data. This 1986 standard (ISO 5349) also gives the set of curves shown in Figure 11.3 that can determine exposure levels likely to cause the first signs of white finger in workers.

The horizontal axis in Figure 11.3 represents vibration acceleration. This is measured as RMS (Root Mean Square) weighted acceleration in m/s^2. RMS is a method of determining average for quantities that fluctuate with time about a central value. Weighting accounts for variation in human sensitivity to vibration of different frequencies. The measured value of acceleration at different frequencies is passed through a weighting filter (see Figure 11.2, Representation of Vibration) to obtain a single number as an overall measure of vibration exposure. According to this frequency-weighting filter people are most sensitive to hand arm vibration in the frequency range of 1/3 octave bands with centre frequency 6.3 to 16 Hz. As frequency increases above this range sensitivity decreases. The standard provides methods for calculating weighted RMS acceleration and equivalent acceleration values where the level of daily exposure varies with time.

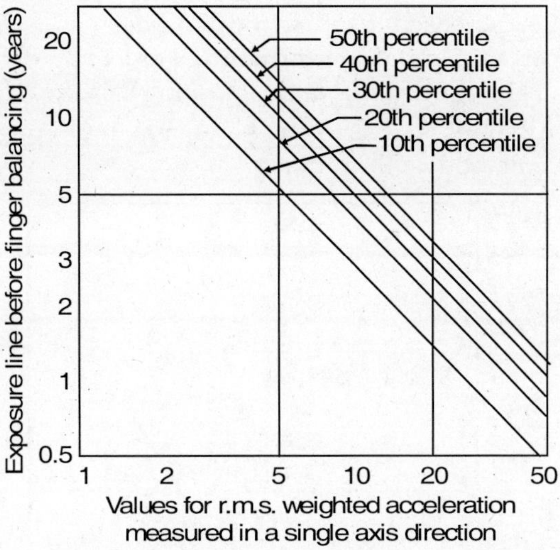

Fig. 11.3. Curves for exposure times of percentiles of population groups (ISO 5349.2) to suffer mild effects on tip of finger (see Stage 1, Table 11.2)

Figure 11.3 can be used to assess the long-term effect of 4-hour per day exposures to hand-arm vibration. For example, the standard predicts that exposure to 50 m/s^2 vibration acceleration will cause 50 per cent of exposed workers to reach stage 1 of Raynaud's phenomenon of occupational origin in about 1.2 years. At a vibration acceleration of 5 m/s^2, the standard predicts that it would take about 14 years for the same percentage of workers to reach stage 1. The data for which the curves were generated are limited, so they should not be used if exposure is greater than 50 m/s^2 or if the exposure duration exceeds 25 years.

11.10. STANDARDS OR GUIDELINES FOR EXPOSURE TO WHOLE-BODY VIBRATION

The standards and guidelines concerning whole-body vibration are designed to reduce vibration to a level where most workers can perform job tasks without discomfort.

The ISO standard gives three different types of exposure limits:

- a reduced-comfort boundary
- the fatigue-decreased proficiency boundary
- an exposure limit

The reduced-comfort boundary is for the comfort of people travelling in aeroplanes, boats, and trains. Exceeding these exposure limits makes it difficult for passengers to eat, read or write when travelling.

The fatigue-decreased proficiency boundary is a limit for time-dependent effects that impair performance. For example, fatigue impairs performance in flying, driving and operating heavy vehicles.

The exposure limit is used to assess the maximum possible exposure allowed for whole-body vibration.

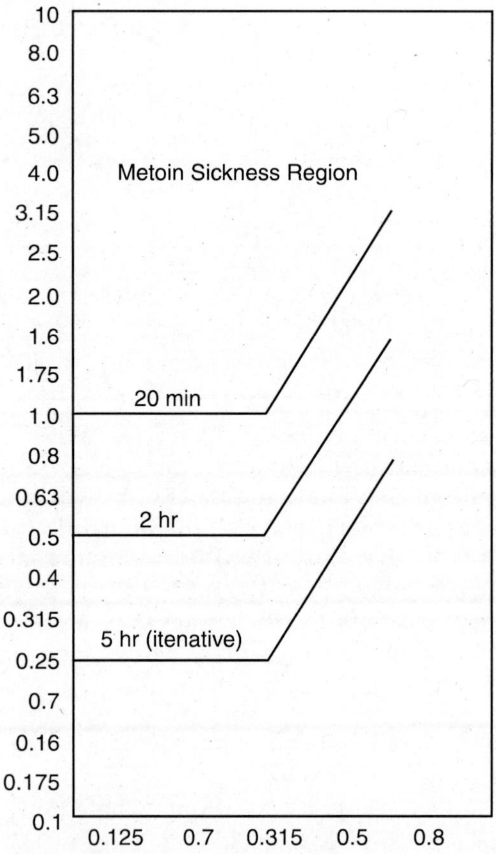

Fig. 11.4. ISO 2631 "severe discomfort boundaries"

A separate set of "severe discomfort boundaries" is given for 8-hour, 2-hour and 30-minute exposures to whole body vibration in the 0.1 Hz to 0.63 Hz range. As with all standards, it is important to read and understand all the information before applying it in the workplace.

These exposure limits are given as acceleration for one third octave band frequencies and three directions of exposure - longitudinal (head <-> toe) and transverse (back <-> chest and side <-> side). The exposure limit is the lowest for frequencies between 4-8 Hz as the human body is most sensitive to whole-body vibrations at these frequencies.

It is important to remember that people vary in their susceptibility to effects of exposure to vibration so the "exposure limits" should be considered as guides in controlling exposure: they should not be considered as an upper "safe" limit of exposure or a boundary between safe and harmful levels.

IMPORTANT QUESTIONS

1. Define vibration. What are the characteristics to describe a vibration?
2. Explain the effect of vibration on human body.
3. Define 'Resonance'.
4. What are the symptoms of 'HAVS?
5. What are the limits of vibration exposure?
6. How vibrations can be measured?
7. Discuss the methods for controlling exposure vibration.
8. Discuss the guidelines for exposure to whole-body vibrations.

CHAPTER 12

Lighting for Working

12.1. INTRODUCTION

Light and vision have their influence on work performance. A well designed lighting scheme increases individual and group productivity and reduces accidents at work. It also reduces rejections in processes and fatigue of individuals. It gives a feeling of well being to the workforce as well.

On the other hand, a poor lighting can cause (*a*) accidents, (*b*) in comfort to workers, (*c*) Defects in the eyesight of the workers in the long run, generates unpleasantness and leads to less output, more fatigue, more accidents and more rejections. For this reason, lighting in industry is an important area for action for the management as well as the safety professionals.

12.2. PURPOSE OF LIGHTING

The purpose of industrial lighting is to provide energy efficient illumination in quality and quantity sufficient for safety and to enhance visibility and productivity within a pleasant environment. Industry encompasses seeing tasks, operating conditions and economic considerations for a wide range. Visual tasks may be extremely small or very large; dark or light: opaque, transparent or translucent; secular or diffuse surfaces and may involve flat or contoured shapes. With each of the various tasks conditions, lighting must be suitable for adequate visibility and productivity in developing raw materials into finished products. Physical hazards exist in manufacturing processes and therefore, lighting must contribute to the utmost as a safety factor to increase the speed of vision. Lighting must serve not only as a production tool and as a safety factor, but should also contribute to the overall environmental conditions of the work space.

12.3. COMMON TERMS USED IN LIGHTING

Some of the terms in illumination engineering are confusing to average person. A few of these terms are:

(*a*) Radiant flux (R) is the total energy radiated by a source per sec. Its SI unit is watt (W)

(*b*) Luminous flux (ϕ) is the term used to represent the total light available. Lumen (l_m) is the unit for luminous flux. The units lumen and watt have same dimension as they both represent energy per second.

(*c*) Luminous efficiency (η) is ratio of luminous flux emitted by a source to the total radiant flux *i.e.*,

$$\eta = \frac{\phi}{R}$$

Its units is lumen/watt.

(*d*) Luminous intensity (*L*) is the term used to measure energy emitted by a light source in a particular direction. Its unit is candela.

$$L = \frac{\phi}{4\pi}$$

(*e*) Candela (*cd*) is the luminous intensity, in a given direction, of a source that emits monochromatic radiation of frequency 540×10^{12} hertz and that has a radiant intensity in that direction of 1/683 watt per steradian.

(*f*) Illumination (*I*) is the term used to represent the light falling on a surface. Technically, it will be more correct to say that it is density of light flux incident upon a surface

$$I = \frac{\phi}{4\pi r^2}$$

(In case of point source at a distance *r*)

$$= \frac{L}{r^2}.$$

lllumination has two units.

The Foot cande — in FPS unit — which is equal to the illuminantion on a surface of an area 1 square foot, created by 1 lumen of light flux.

Lux — Metric system — is the illumination on a surface 1 square metre in area, created by 1 lumen of light flux.

1 lux = 10.76 foot candle.

(*g*) Luminance is the brightness of a surface. It is different from illumination. Luminance of a surface depends upon the light falling on the surface and also the ability of surface to reflect light. With the same amount of light, a black piece of cloth will have considerably less brightness than a mirror. Brightness is measured in candler per square meter or in sulb (1 sulb = 10000 candela/sqm)

Example 12.1. A lamp emits 60 W of radiant flux distributed uniformly in all direction. The luminous efficiency is 60 lumen/watt. Calculate

(*i*) Luminous flux of the lamp.

(*ii*) Luminous intensity

(*iii*) Illuminance on a surface 2 m apart.

Solution:

$$\eta = \frac{\phi}{R}$$

$$\text{Luminous flux } (\phi) = \eta \times R$$

$$= 60 \times 60 = 3600 \text{ lumen.}$$

$$\text{Luminous intensity } (L) = \frac{\phi}{4\pi}$$

$$= \frac{3600}{4 \times 3.14} = 286.6 \text{ candela}$$

$$\text{Illuminance } (I) = \frac{L}{r^2}$$

$$= \frac{286.6}{2^2} = 71.65 \text{ lux.}$$

Example 12.2. The luminous intensity of a high source is 400 candela. Find the intensity of illumination on a surface placed at a distance of 2 m.

Solution:

$$I = \frac{L}{r^2} = \frac{400}{2^2} = 100 \text{ lux.}$$

Example 12.3. A light source emits 30 Joules of energy in 10 seconds. What is the radiant flux of the source?

Solution: Radiant flux $(R) = \frac{\text{Energy}}{\text{Time}} = \frac{30}{10} = 3$ watt.

Example 12.4. An illumination of 100 lumen/m² is provided in a seminar hall 40 m × 10 m. Assume coefficient of utilization as 0.4 and depreciation factor 0.8 and efficiency of lamp is 14 lumen/watt. Calculate the number of lamps (1 lamp = 500 watt).

Solution:

$$I = 100 \text{ lumen/m}^2$$

$$\text{Area, } A = 40 \times 10 = 400 \text{ m}^2$$

$$\phi = I \times A$$

$$= 100 \times 400 = 4 \times 10^4 \text{ lumens}$$

$$\text{Lamps output required} = \frac{4\times10^4}{0.4\times0.8} = 1.25 \times 10^5 \text{ lumens}$$

$$\eta = 14 \text{ lumen/watt}$$

$$\text{Power} = \frac{1.25\times10^5}{14} = 8928 \text{ W}$$

$$\text{Number of bulbs required} = \frac{8928}{500} \approx 18 \text{ bulbs.}$$

12.4. BASIC CONSIDERATIONS IN LIGHTING

It is a common misunderstanding that providing good lighting conditions means only increasing the illumination level. Sufficient light to perform the task comfortably, of course, is important, but there are many other factors which need to be considered in the study of lighting.

(i) Light and Visibility

Lighting is for seeing. Visibility or the case of seeing depends upon three interrelated factors:

(*a*) **The size of the object.** The larger the size, the easier it will be to see. Inspecting a watch is more difficult compared to examining a clock, in terms of eye strain, with the same light.

(*b*) The amount of light falling on the object. Correct levels of lighting makes it easier to see the task.

(*c*) The contrast between the object and the surroundings. It will be easier to see a black spot on a white shirt than on a black suit.

(ii) Age and Light Needs

As the age advances, the light requirements of individuals increase. A person aged 60 will need approximately 5 times the light required by a person aged 40 for the same task under the same conditions. A boy aged 10 will need 1/3 of the light needed by the latter. This is probably the reason why a youngster can easily spot a needle fallen on the floor much faster than a middle aged person.

Further, as age advances the minimum distance at which things are clearly visible moves away from the eyes.

(iii) Accommodation

The human eye can adjust itself to many changes in the tasks that it performs. It can, for instance, adjust to the distance. One moment a person is reading, next moment looking at a car coming up the hill and then looking at the clock on the wall. All the three tasks are performed by the eye with ease. While moving from a dark room to a bright room or vice versa, the eye adjusts itself to the change in level of illumination in a matter of seconds. It can function in dull moonlight and bright sunlight. This ability to self adjust, called accommodation, is due to the fact that the retina, the pupils and the muscles of the eye are capable of adjusting to the need and controlled by the brain. However, if the eyes are forced to make use of this ability continuously, it will overwork the muscles and lead to eye strain.

Typical situations where eye strain can occur are:

A machine operator faces the window. His eyes travel form the job to the window and back many times and by the end of the day, he can feel eye strain and possibly headache. A person who is required to read a number of meters in a row, and one required to see a luminous display in the wall and write it down continuously are likely to feel the same effect.

The visual effort required should be studied for each job and wherever possible attempts made to reduce the strain in eyes by not forcing it to accommodate wide variations continuously. In the first case, providing a window curtain or turning the machine around so that the operator does not have to face the window could solve the problem.

(iv) Use of Colours

Colours are used in industry to fulfil two needs — one is to enable objects or surfaces to be distinguished from one another. The other is to "give a good look" i.e, to provide acceptability in terms of psychological effects on people.

Selection of colour for an object or surface depends on needs predominant. In the case of an ON and OFF switch for a machine the predominant need is to distinguish one knob from the other and both knobs from other machine parts. A bright contrast will be the ideal choice here like bright green and bright red for the knobs and a light back-ground for the surroundings.

But when it comes to painting of walls or ceilings or machines, the predominant need is to give an optimum psychological environment, though distinguishing one object or surface from another also figures. (Say, a door from the wall). Light pastel shades may be more suitable for this purpose

Different colours have different effects on people in terms of psychological acceptance. Blue and green have a soothing effect whereas orange, yellow and brown are stimulating. Red and violet are "eye catching" but on large surfaces, they have a disturbing and sometimes alarming effect.

The type of work being carried out also needs to be considered in the selection of colours. For dull, monotonous tasks, stimulating colours are recommended. Soothing colours around can give some amount of relief to persons engaged in work involving concentration and mental fatigue.

Light colours on the wall reflect more light and make the workplace look brighter. An artistic selection of colours, taking into account, the factors mentioned above and the people can make the workplace a happier place to stay.

(v) Glare

Apart from causing strain, glare can sometimes blind the operator for a few seconds. The person himself may also close the eyes by reflex. While operating or working near machines both cases can lead to severe accidents as well. So the causes of glare should be eliminated as far as possible.

Some of the sources of glare in workplaces are:

(*a*) Windows admitting or reflecting sunlight.

(*b*) Highly polished surfaces like large plated sheets in process or bright inspection blocks.

(*c*) Unscreened light sources like bulbs or tubes, in the visual field. (Visual field is the area that is visible when the head and eyes are kept still). This is a problem particularly in large sheds, where naked bulbs/tubes in one end are directly visible from the other end.

(*d*) In cases of workplaces, situated near main roads, headlights of cars at night also can cause glare.

(*e*) If the light sources are not carefully located light rays falling on the workbench itself can be directly reflected to the eyes of the operator.

Glare can be avoided by (a) eliminating large reflecting surfaces from the direct path of light and (b) by-suitably adjusting the layout such that reflecting surfaces that cannot be eliminated are kept away from the visual fields of persons working.

12.5. LEVELS OF ILLUMINATION

Technically, one can see even with very low levels of illumination. But in this case, the human efforts to be put in to see and work will be too much to be justified from productivity considerations. The recent trend, therefore, is to provide sufficient illumination to perform the task well without fatigue and least chances of error.

The standards for illumination levels followed in different countries vary to some extent. It will be advisable to consult the national standards on illumination (and in its absence the international standards) for information relating to illumination levels required for specific tasks. Leading Lighting Engineering Firms also provide information service in this respect.

The level of illumination to be provided also will depend upon:

- Specific task requirements.
- Reflection of light from walls and ceilings.
- Age of the employees, and
- Prevailing daylight conditions

Obviously the investment required and the cost of maintenance also are not less important.

The values of illumination levels recommended by IES code for some selected tasks are given below:

Approximate Guide to illuminance Levels.
Source IES Code 1977

Typical Task or Interior	**Standard Service Illuminance (Lux)**
Storage areas and plant rooms with no continuous work	150
Casual Work	200
Routine Work, Rough Machining and Assembly	300
Routine Work, Offices, Central Rooms, medium machining and assembly	500
Demanding Work, Deep Plan, Drg Office, Inspection of medium machining.	750
Fine Work Colour discrimination textile processing, fine machining	1000
Very Fine Work, Hand engraving, Inspection of fine machining or assembly.	1500
Minute work, Inspection of very fine assembly.	3000

12.6. SOURCES OF LIGHT

Daylight is the natural source of light and has the great advantage that it does not cost anything. But daylight is not available for more than 8 hours in a day and as such, one has to look for other sources as well. Electric light is the next and perhaps the only alternative. It is the common sources of light used extensively.

Daylight

Apart from the cost, day light has another plus factor–its excellent colour rendering qualities.

Daylight can be admitted to the workplace through the windows on the sides or inlets provided on the roof. The problems associated with the admission of sunlight are:

(*a*) Lighting by daylight alone cannot be depended upon, due to variation in lighting conditions during different hours of the day and from season to season. Therefore, artificial lighting is to be provided and the use of daylight can only be to supplement or substitute them during day time. Even this is a great advantage in terms of power savings and life of artificial fittings.

(*b*) Admission of daylight through windows is associated with chances of glare, especially in tropical countries. This problem can be overcome by limiting the number of windows/inlets admitting sunlight and positioning them away from the visual field of persons.

(*c*) Unplanned admission of sunlight results in dark and bright spots. This can be controlled by careful positioning of the inlets taking into account the travel of sun and diffusing the light admitted (For example, by using translucent glass for windows and inlets).

Artificial Lighting

The choice available are from ordinary electric bulbs (Tungsten filament) Tubelights (Fluorescent tubes) and Mercury vapour lamps.

Ordinary electric bulbs are not very suitable for industrial lighting because of their many disadvantages. These include:

(*a*) Low light output per watt consumed.

(*b*) They radiate more heat.

(*c*) Unless encased in a shield, they create glare. With a shield light output is further decreased.

The one advantage which ordinary bulb has, is that they need minimum initial investment. They are suitable for portable applications and providing supplementary lighting (as in the case of an extra, adjustable lamp for the machine).

Fluorescent lamps are being increasing if used in industry. They have their own advantages and disadvantages, but the former outnumber the latter. Advantages are;

(*i*) High light output per unit consumed.

(*ii*) They give lighting conditions that resemble daylight.

(*iii*) Least glare–glare can be totally eliminated by the use of levers and diffusing shields.

(*iv*) Elegant to look at.

Fluorescent lamps have two disadvantages;

(*i*) They flicker and have a 'stroboscopic' effect on rotating parts

(*ii*) They need higher initial investment.

The third choice is the Mercury vapour lamp. The light output per unit consumed is favourable though less than fluorescent lamps. These lamps come in high ratings which is useful in certain applications (e.g. High bay lighting). Their disadvantages are:

(*a*) Very poor colour rendering qualities (this can be overcome to a limited extent by having ordinary bulbs of some rating nearby)

(*b*) They create glare.

The fourth choice is the Metal Halide lamps. These lamps use a combination of several metals and have high light output per unit consumed than mercury vapour lamps. Its light is described as cool blue in perception with a more noticeable colour shift with increasing hours of operation.

These lamps have good colour rendering qualities and high efficacy. These can be used where large open areas are required to be lit. This is the preferred light source for indoor industrial uses, monuments, bridges, enhancing architect, usal features and sports arenas. It is also used in warehouses, street lighting and some retail applications.

High Pressure Sodium lamp, commonly referred to by its HPS initials. Traditionally it was associated with its yellow or orange color. Newer and more expensive lamp designs have high color rendering value upto 85 and its color normally has a very warm golden white tone. It is the preferred choice where maximum energy efficiency, high light output and very long lamp life is required. High color rendering value can be obtained at a considerable drop in efficacy.

This lamp is principally specified for roadway, street and parking lighting, as well as indoor industrial lighting where color rendition is not critical. Warehouse lighting is also a common application.

12.7. LIGHT CONTROLS

There are several means available to control the lighting levels. We will concentrate only on two technologies motion or occupancy sensors and dimming systems.

Motion or occupancy sensors are used to automatically switch lights. Lighting is either switched off or switched down when these are newer sensors that combine passive infrared, ultrasonic and photocell technologies for more complex applications. There are many sensor brands available and they do not necessarily operate alike. Sensor location is critical to a successful application. Incandescent lamps are the easiest to control and switch because they can be instantly switched on or off. Fluorescent lamps require more complex controls, but they can also be instantly switched on or off. High intensity discharge lamps can be switched off easily; however after being switched on, it takes several minutes for them to attain full light output. This makes it necessary to use either an auxiliary quartz lamp or a separate lamp for emergency purposes.

A bi-level switching system can be used for high intensity discharge fixtures to drop their power consumption to approximately 50 per cent of nominal capacity. With this system, the lamp can be instantly switched up to 100 per cent power when needed. This bi-level system has proven very cost effective in warehouses, where some aisles are occupied only a small percentage of the time.

Dimming systems are available for all lamp types. Incandescent lamps are the easiest ones to dim. Fluorescent lamps also lend themselves easily to dimming but require more expensive controls and dimmable type ballasts.

12.8. MAINTENANCE STANDARDS

1. Fluorescent fittings equipped with electronic ballasts (and connected to dimming control equipment), must be treated as systems, rather than individual fittings. Tube replacement should be undertaken on a "whole room" basic at periodic intervals, as demanded by effective tube life. Universal policy is to relamp each

theatre during the summer semester break. Tubes should not be replaced on an individual basic, as new tubes will exhibit totally different dimming characteristics to used tubes.

2. Fittings must be planned to allow simple replacement. In large theatres with high ceilings, access is required above the ceiling on permanent catwalks.
3. Acrylic diffusers, with built-in silicon gasket, to prevent loss of gasket during lamp.
4. Stainless steel diffuser clips to be used for ease of maintenance.
5. A minimum of IP 65 for both lamp and control gear compartment for high maintenance factors.
6. Gear tray and diffuser to be attached to body via a lanyard for ease of maintenance.
7. Provision for photocell that can be easily replaced.
8. Regular cleaning and prompt replacement of lamp outages is essential in any well-operated industrial lighting system. Analyse luminaire construction and reflector finish and also to make provisions for maintenance access so the system can be properly serviced.

12.9. DESIGN CONSIDERATIONS FOR LIGHTING

The designer of an **industrial lighting** system should consider the following factors as the first and all-important requirements of good planning. 1. Determine the quantity and quality of illumination desirable for safety of personnel, the manufacturing processes and the environment. 2. Select **lighting** equipment that will provide the quantity and quality requirements by examining photometric characteristics, and mechanical performance that will meet installation, operating and actual maintenance conditions. 3. Select and arrange equipment so that it will be easy and practical to maintain. 4. Balance all of the energy management considerations and economic factors including initial, operating and maintenance costs, versus the quantity and quality requirements for optimum visual performance. The choice of the electric distribution system may affect overall economics. The use of daylighting should be considered for area **lighting** in all industries.

Factors of Special Consideration

1. **Lighting and Space Conditioning.** With the use of higher illuminances, it is often practical to combine the **lighting**, heating,

cooling and atmospheric control requirements in an integrated system. The **lighting** system can often provide most of the energy during the heating period. When cooling is required much of the **lighting** heat can be removed by the air exhaust system.

2. **High Humidity or Corrosive Atmosphere and Classified Location Lighting.** Enclosed gasketed luminaries are used in non-classified areas where atmosphere contain nonflammable dust and vapours. Enclosures protect the interior of the luminaries from conditions prevailing in the area. Steam processing plating areas, wash and shower rooms, and other areas of unusually high humidity are typical areas that require enclosed luminaries. Severe corrosive conditions necessitate knowledge of the atmospheric content to permit selection of proper material for the luminaire. Classified locations are areas where atmospheres contain inflammable dusts, vapors or gases in explosive concentrations. They are grouped by the *National Electrical Code* on the basis of their hazardous characteristics, and all electrical equipment must be approved for use in specific classes and groups. Luminaries are available specifically designed to operate in these areas.
3. **Abnormal Temperature Conditions.** Low ambient temperatures must be recognized as existing in such areas as unheated heavy **industrial** plants, frozen food plants and cold storage warehouses. Equipment should be selected to operate under such conditions and particular attention should be given to lamp starting and light output characteristics, if fluorescent equipment is considered. With high intensity discharge equipment, temperature variation has practically no effect on light output, but the proper starting characteristics must be provided. With incandescent filament lamp equipment neither the starting nor the operation is a problem at low temperature. Abnormally high temperatures may be common in truss heights in foundries, steel mills, forge shops, etc. Caution should be observed in selecting **lighting** equipment for mounting in such locations, it is particularly important to consider the temperature limitations of fluorescent and high intensity discharge ballasts under such conditions. The reduction in fluorescent lamp output at high operating temperatures should be recognized.
4. **Maintenance.** Regular cleaning and prompt replacement of lamp outages is essential in any well-operated **industrial lighting** system. It is important for the **lighting** designer to analyze luminaire construction and reflector finish and also to make provision for

maintenance access so the system can be properly serviced. Another point that should be considered is that it may often be necessary to do the servicing during the plant operating hours.

Supplementary Lighting in Industry

Difficult seeing tasks often require a specific amount of quality of **lighting** which cannot readily be obtained by general **lighting** methods. To solve such problems supplementary luminaries often are used to provide higher illuminances for small or restricted areas. Also, they are used to furnish a certain luminance, or colour, or to permit special positioning of light sources to produce or avoid highlights or shadows to best portray the details of the task. Because supplementary **lighting** can be specified, it is necessary to recognize the exact nature of the visual task and to understand its light reflecting or transmitting characteristics. An improvement in the visibility of the task will depend upon one or more of the four fundamental visibility factors—luminance, contrast, size and time. Thus, in analysing the problem, the designer of the **lighting** may find that seeing difficulty is caused by insufficient luminance, poor contrast (veiling reflections), small size, or that task motion is too fast for existing seeing condition. The planning of supplementary **lighting** also entails consideration of the visual comfort of both those workers who benefit directly and those who are in the immediate area. Supplementary equipment must be carefully shielded to prevent glare for the user and his associates. Luminance ratio should be carefully controlled. Ratios between task and immediate surroundings should be limited. To attain these limits it is necessary to coordinate the design of supplementary and general **lighting**.

12.10. MEASUREMENT OF LIGHT

Measurement of illumination levels becomes necessary to (a) check the prevailing levels with levels planned (b) study the lighting conditions to plan improvements and (c) ensure the reliability of lighting schemes. The illumination level can be directly measured with the help of a direct reading instrument commonly known as light meter or illumination meter. These meters are handy and consist of a light sensing network using photocells. Light energy is converted into electric energy by the network which actuates the meter. The meter in turn, is calibrated to read illumination levels in Lux or foot candles. The light meter also incorporates a battery and a means to check the battery.

12.11. PRINCIPLES OF ILLUMINATION

1. Uniform distribution of light throughout the work-place without dark spots and bright pools.
2. Absence of contrasts that lead to visual fatigue.
3. A reasonable harmony between day-light conditions and artifical lighting conditions.
4. Overall lighting conditions that are pleasant and soothing to the eye.
5. Sufficient quantity of light available for individual tasks to be performed without eye strain and operational errors.
6. Judicious planning of the brightness levels and color schemes to provide proper illumination. As for example the value of reflection factor in case of walls and ceilings could be as low as 0.1 for dark colours and as high as 0.7 for white colours.
7. Proper maintenance of artifical lighting system as well as proper maintenance of the interior and the exterior of the building help a lot in minimising illumination problems. Since the effectiveness of luminaires depends upon the utilization factor and maintenance factor, care must be taken to choose the right luminance as well as proper upkeep.
8. The planning of an industrial complex must be judicious enough so that no confusing shadows or sudden change of illumination level take place. Proper hierarchy of illumination must be maintained so that workers negotiating different levels of illumination can move and work with ease.
9. If required directional lighting should be provided to assist in a typical visual task to be performed. As for example, for buffing and polishing in plating shops directional lighting may be used since the general level of illumination in the shop could be inadequate.
10. Glare whether direct or reflected should be controlled.
11. Maintenance of proper illumination level appropriate to the visual tasks to be performed.

✎ IMPORTANT QUESTIONS

1. What is purpose of industrial lighting?
2. What are benefits of good illuminations?

3. Differentiate between natural light and artificial light.
4. What are various types of artificial lighting? State its objectives.
5. What are principles of good illumination?
6. What are various standards of illumination recommended in an industry?
7. Discuss the maintenance standards needed for lighting and colour so that healthy and conducive environment can be generated for higher productivity.
8. What are the considerations to design for lighting industrial areas?

CHAPTER 13

Industrial Waste

13.1. INTRODUCTION

Industrial waste may be defined as unwanted materials produced in or eliminated from an industrial operation. These wastes are discharged from factories, processing plants, wholesale establishments, assembling plants or shops and garages, such as paper, cardboard, cartons, food processing wastes, cinders and ashes, lumber scraps, rawdust excelsior, shavings, floor sweepings, metal scrap and shavings, glass and other waste products. Industrial waste typically contains somewhat higher levels (upto four times) of contaminants (such as heavy metals and human-made chemicals) than municipal solid waste and needs to be managed with environmental controls appropriate to the specific wastes being landfilled. These are categorised under a variety of headings, such as liquid wastes, slodge, solid wastes and hazardous wastes.

13.2. TYPES AND SOURCES OF INDUSTRIAL WASTE

1. **Liquid waste:** It can be defined as waste water resulting from industrial processing operations. Water is essential raw material in almost all manufacturing plants and it gets contaminated during its usage. Practically all the "wet-processing" industries such as those engaged in manufacture of food products, fruits, vegetables processing and canning, Sugar industries, dairy industries, paper and pulp manufacturing, number of chemical industries, laundries consume large quantity of water and produce waste in large quantity.
2. **Solid waste:** The solid waste may result from wood working industries (raw dust), the fragments of metal fabrics, plastics etc., resulting from machining and assembly operations. These wastes are usually salvaged as far as possible and remainder is dumped. Engineering and public health problems arise in their disposal.

   The solid wastes frequently enter the liquid waste stream either accidently or by design – for example, metal chips, mineral fins, raw dust, slossies from chemicals etc.

3. **Gaseous waste:** In some industries nuisance or poisonous gases, fumes and dusts are produced–for example, melting furnaces, foundry, chemical industries, welding, smithy refineries, power stations (thermal) etc. These are disposed into atmosphere unless special treatment operations are provided. The problem of atmospheric pollution (air pollution) is a result of such gaseous waste.

13.3. HARMFUL EFFECTS OF INDUSTRIAL WASTE

1. Many Industrial wastes are discharged at high temperature such as condenser water from power plant, waste from paper mills, waste from textile industry, coke plants and many chemical plants. Sudden discharge of high temperature waste may result.
   (*a*) Killing of fish in streams.
   (*b*) Reduction of dissolved oxygen at high temperature creates danger to aquatic life.
   (*c*) the formation of inflammable gases such as methane and toxic gases such as hydrogen sulphide are dangerous to human life.
2. Suspended solid produced by waste form tanneries, starch and corn products, textile mills impair normal aquatic life, form benthal deposits, and may get decomposed.
3. Industrial waste in stream may carry chemical substances such as phenolic compound, hydrogen sulphide. It causes tastes and odour and makes the water unsuitable for drinking.
4. Discharge of petroleum products and such other liquids insoluble in water often results in oil film. Heavy film of petroleum oils and inflammable organic liquids results in fire hazard. The oil film also prevents absorption of oxygen from atmosphere and, therefore, it is dangerous to aquatic life.
5. The acidic waste may origin from wet processing industries such as food industries, bleacheries, textile mills, coalmines, iron and steel industries, electroplating and chemical industries which are harmful to natural streams and severage streams in many ways such as:
   (*a*) Kill the fish
   (*b*) It may cause corrosive action to metals and other construction materials
   (*c*) Makes the water unsuitable in drinking, swimming, navigation etc.

6. The floating solids produced by breweries, distilleries, paper-mills cover appreciable stream surface and may retard oxygen absorption by water.
7. Organic substances when dissolved in water even in very small quantity result in formation of bubbles and accumulation of froth due to aeration or agitation. These material
   (*a*) affect absorption of oxygen by water
   (*b*) reduce light penetration and are harmful to aquatic life.
   (*c*) some organic materials are toxic.
8. Inorganic materials may give rise to various problems such as hardness and corrosion.
9. Colour and turbidity create unaesthetic conditions in water bodies. As a result oxygen availability and photosynthesis are also affected.
10. Heavy metals, cyanides and toxic substances are harmful to living groups.

13.4. CONTROL OF INDUSTRIAL WASTE

It is not possible to eliminate the industrial waste but if efforts are made then these can be reduced to a great extent. These wastes can be reduced by the following means:

1. Improving the product design
2. Using safe and efficient material handling system.
3. Adopting proper inspection procedure and effective quality control system to minimise the wastage
4. Maintaining machines properly to reduce the casts associated with the down time.
5. Improving material cutting layouts.
6. Changing the manufacturing processes example use die casting or shell moulding instead of using sand casting method to avoid machining operations, and reduce wastage of material.
7. Instead of close tolerances propose moderate tolerances which will help to reduce number of resections.
8. Resharpening of tools after getting dull.
9. Proper control of temperature and humidity in store rooms so as to reduce shrinkage.
10. Changing purchase specifications to avoid losses due to left overs.

In all industrial processes there is loss of raw materials and other material throughout the manufacturing operations inspite of all the efforts mentioned above. These materials become the source of pollution and as well as economic loss to the manufacturer. Hence the recovery of the material should be thoroughly investigated and recovery methods should be implemented. The recovery from waste results in economy and reduces load on waste treatment plant.

13.5. WASTE RECOVERY

➢ Water Recovery

Water is an expensive raw material in many process industries and if it can be reused in the processes substantial economy is achieved. Reuse requires partial purification of water and if it can be achieved at reasonable cost, reuse is beneficial. The water from waste stream can be used for downgraded reuse. For example, contaminated effluents from coke and gas manufacture can be reused for quenching of coke.

➢ By-products Recovery

Along with the economy of the water reuse there may be salvage of heat and chemicals contained in water. Coal is washed in order to remove impurities from mechanical mining. The waste water contains suspended coal particles. The fine coal is easily recovered and converted into marketable form. The water is also suitable for reuse. In metal plating industry, considerable quantities of metal salts are lost in rinse water: for example, chromic acid in the rinse following chromium plating operation. Partial recovery if can be achieved by installing a rinse tank immediately after plating tank, using water from this rinse tank to replenish the losses in plating tank. Recovery of metal salts by evaporation of rinse water is many a times economical.

13.6. TREATMENT OF INDUSTRIAL WASTE

The purpose of treatment of industrial waste is to recover the materials lost in the industrial waste for their reuse. The waste treatments may be classified on the basis of individual plant requirements. The waste treatment may be classified as (a) Physical treatment (b) Chemical treatment (c) Biological treatment.

(*a*) Physical Treatment: The physical treatment is least expensive and requires simple equipment. It includes:

(*i*) Sedimentation

(*ii*) Floatation

(*iii*) Filtration

(*iv*) Heat transfer operation

(*i*) Sedimentation: It is used to remove settleable particle. It is operation by which suspended particles in liquid settle by gravity to the bottom of the tank, from which they may be removed as sludge. At the same time, lighter solids and oils float on the surface of liquid and can be separated as scum. Sedimentation is affected by quiet standing of liquid or by very slow streamlined flow through a tank or channel with minimum turbulence. The following aspects are considered in design and construction of setteing tanks:

Concentration of settleable grit and solids to be removed.

Quantity of waste to be treated.

Oil and grease separation requirement.

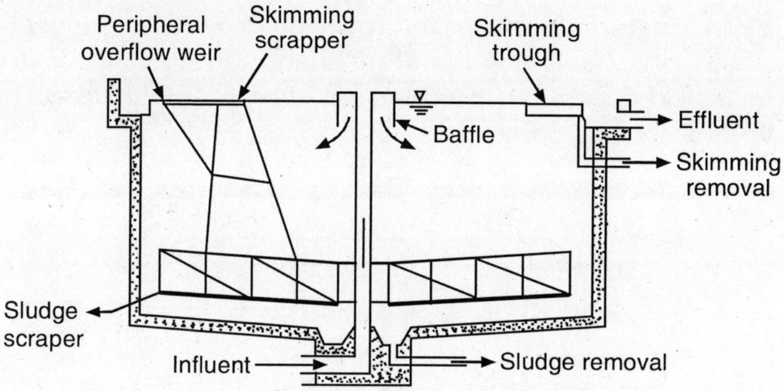

Fig. 13.1

(ii) **Floatation:** In this process air bubbles are introduced into waste water, which gets attached to the suspended solids. This reduces specific gravity of suspended solids, thus permitting their rise to the surface. The sludge in the form of scum collected on the surface is then skimmed. This process is also useful for removing oil and grease in waste water.

(iii) **Filtration:** Filtration is defined as the mechanical separation of suspended particles by passing the liquid waste through porous medium which retain the solids. This process is rapid and requires less space but this is expensive process. The basic requirement of filtration device are:

A porous medium such as sand bed, porous stones, screen, paper or cloth with adequate support.

Space for accumulation of solid as filter cake.

Provision for removal of filtrate.

Supply of forced liquid under pressure or arrangement of withdrawing filtered liquid by vacuum.

Means for washing deposited solids.

(*iv*) **Heat transfer operations:** Heating may be necessary for sterilisation of waste or killing of bacteria. Heating is sometimes employed to drive off volatile gases from the waste, to dry sludges or concentrations from primary treatment.

(*b*) **Chemical Treatment:** It includes

(*i*) PH Adjustment

(*ii*) Oxidation

(*iii*) Chlorination

(*iv*) Reduction

(*i*) **PH Adjustment:** One of the common types of chemical treatment of industrial waste is PH adjustment. The Ph scale is considered to extend from zero (the utmost acidity) to 14 (the upmost alkalinity) with a normal point at PH_7.

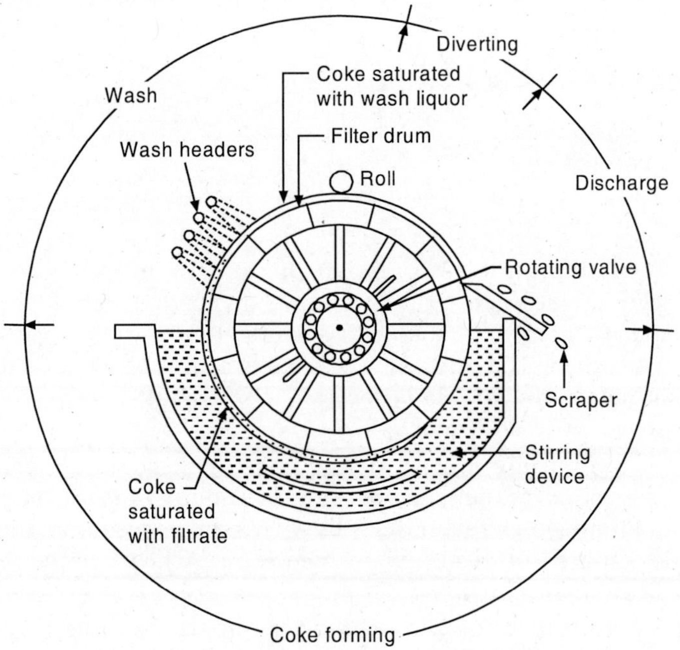

Fig. 13.2

Acidic wastes are harmful to natural streams and severage streams in many way, such as it may cause corrosive action on metals and other construction materials, kill the fish and make the water unsuitable for drinking , swimming, navigation etc. PH adjustment consists of removal of excess acidity or alkalinity. It is called neutralization and is accomplished by using suitable reagents.

Neutralization of Acidic Waste: Acidic wastes have their origin in many industrial processes such as chemical plants, coal mines, electroplating shops, iron and steel mills, pulp mills etc. Sodium alkalines such as caustic soda (NaOH) and soda ash (Na_2O_3) are commonly used as basic materials for neutralization of acidic waste. Both are highly soluble in water therefore feeding is convenient and controllable. They also produce less sludge. However, care is taken in handling of these reagents as they affect the skin.

Oxides of calcium and magnesium are considerably cheaper than sodium alkalies and are commonly based. They are not much soluble in water and are to be fed dry or in the form of slurry. Large contact period and agitating is also necessary. They also form appreciable sludge. Common calcium bases used are: Quick lime (C_aO), slaked lime $c_a(OH)_2$, Magnesite ($\mu\, g_o$), Dolomite, $CaCo_3$, $MgCo_3$ etc.

Neutralisation of Alkaline Wastes: Alkaline wastes are produced by many industries such as pulp mills, leather processing, petroleum refining etc. Acids are used as neutralizing agents. For rapid reaction rate and complete neutralization strong acids are used. When reaction rate need not be too rapid and complete neutralization is not necessary cheaper available acid is carbolic acid ($HgCO_3$). The flue gas containing Co is bubbled through alkaline waste for neutralization purpose. The design, construction and operation of chemical plants for PH adjustment depends upon the following factors:

(*i*) Flow rate for waste and its variation.

(*ii*) Acidity or alkalinity and its variation.

(*iii*) Allowable PH range in the effluent.

(*iv*) Kind and degree of agitation.

(*v*) Reagent feeding devices.

A typical pH control system is shown in the figure below:

A pH amplifier
C Recording pH controller
D Diaphragm motor value
G Diaphragm motor operated gate valve
I Immersion type electrode assembly
R Electric pneumatic relay

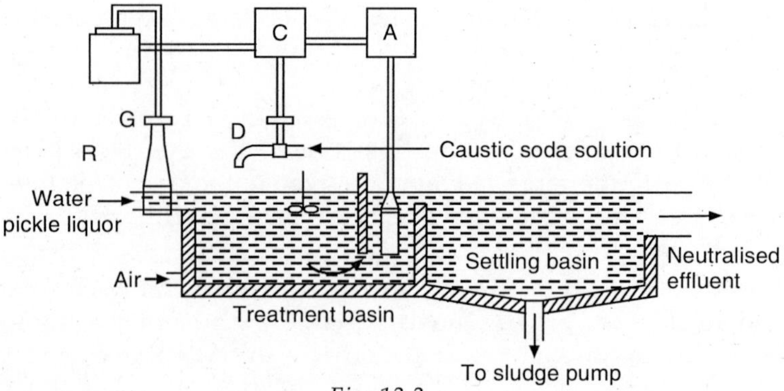

Fig. 13.3.

(ii) Oxidation

The process of oxidation serves the valuable purpose for treating almost all types of industrial wastes. The removal of BOD can be achieved by oxidising the waste constituents. BOD (Bio-chemical Oxygen Demand) is a measure of reducing properties of waste, particularly of its organic constituents. BOD is stated in terms of quantity of oxygen required for the oxidation and destruction of these substances by chemical and bio-chemical reactions. BOD can be defined as "The quantity of oxygen utilized in biochemical oxidation of organic matter in a specified time at specified temperature".

Other specific applications of oxidation include destruction of compounds such as phenols and cyanides. Complete oxidation of organic pollutants leads to formation of harmless materials such as CO_2, nitrites sulphates etc..

The oxidation is achieved by aeration, using the cheapest and abundant source of oxygen in atmosphere. Aeration may be achieved by mechanical agitation of waste and or spray of waste so that intimate contact of waste and oxygen from atmosphere is achieved. Use of compressed air is also made for aeration.

Oxidation is achieved by ozones, specially for treating phenolic and cynide waste, ozone is generated near the place of use by passing air or oxygen through high voltage silent electric discharge.

(iii) Chlorination

Chlorination using liquid chlorine or commercial chlorinated $CaOCl_2$ (calcium oxychloride) or calcium dioxide (ClO_3) is also important process for treatment of various industrial wastes including cynide and phenolic wastes.

(iv) Reduction

Though the process of reduction has limited use, it is important treatment especially in case of chromium waste such as chromates, di-chromates etc. produced as a result of chromium plating, anodising etc. Reduction of chromates takes place effectively in acid solution.

Common reducing agents include ferrous sulphate, metallic iron turnings, metallic zinc or brass, sulphides. SO_2 etc.

Ion Exchange. It involves exchange or replacement of ions present in polluted water by some other ions. A solid ion exchanger in the form of resin when added to polluted water, picks up harmful cations from polluted water and releases different cations in the water. All ionic changers exhibit selectivity towards certain ions over another. The selectivity varies with ionic strength, relative amounts ions, the kind of exchangers, temperature etc. The commonly used exchanger materials are natural zeolites, clay, sulphonamide coal etc., but the synthetic resins are used for large scale ion exchange because of their high capacity and controllable properties.

Ion exchange can be used to remove unwanted ions like nitrate, iron, manganese, ammonium and heavy metals preferentially.

The disadvantage of ion exchange process is that the chemical used for regeneration of exchangers may be expensive, corrosive, or even dangerous and the waste regenerate may present a difficult disposal problem.

(c) Biological Treatment

In includes: (*i*) Aerobic biological treatment
(*ii*) Anaerobic biological treatment

(i) Aerobic biological treatment:

One of the most important methods of treating sewage and organic industrial waste is by action of bacteria and other micro-organism. Under proper environmental conditions, micro-organisms utilize organic waste as food, converting it into simple non-polluting compounds such as CO_2, water, nitrates and sulphates etc. The treatment is usually operated under aerobic conditions, i.e., in presence of free or dissolved oxygen.

Biological processing of sludge accumulated by sedimentation is usually accomplished under aerobic conditions, the process known as sludge digestion.

In order to have aerobic biological activity, it is necessary to supply proper types of microorganisms, suitable food material, source of oxygen and proper environment for growth and reproduction of micro-organisms.

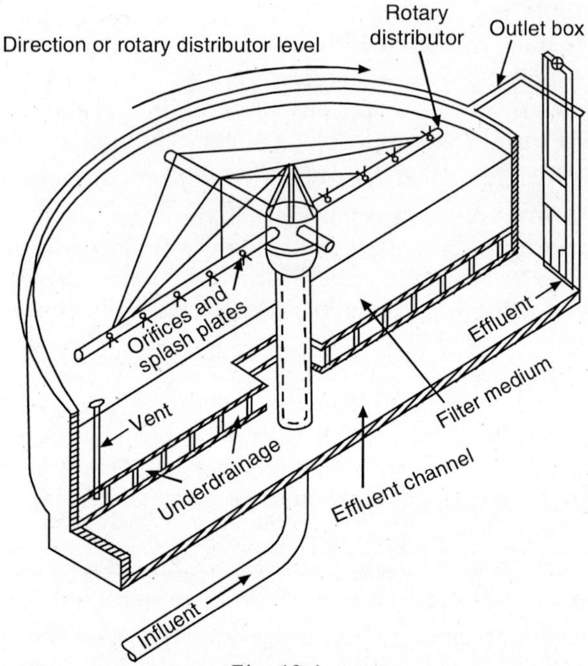

Fig. 13.4.

In industrial waste treatment, the organisms either can occur naturally or can be developed by seeding from other sources. Food is supplied by organic matter in the waste. Proper environmental condition is achieved by controlling temperature. pH and toxic materials by previous treatment. The required air supply is provided by various means.

Trickling filters. Trickling filter is the most popular for aerobic biological treatment. Trickling filters are artificial beds of crushed stones or other porous medium through which sewage or industrial waste is caused to percolate, thus bringing intimate contact with air for active biological growth.

The micro-organisms develop spontaneously from most of the wastes on the surface of the stone. The principal action of trickling filter is decomposition of complex organic molecules present in the waste by hydrolysis and oxidation under aerobic conditions into simpler compounds. The air required is supplied by dissolved oxygen also due to the fact that air moves through the filter under the influence of temperature differences between liquid and atmosphere. If necessary forced circulation may be provided.

Trickling filter bed is usually 2 to 2.5 meter in depth having circular or rectangular shape. Crushed stone is most common filter medium, however slag unthracite, coal are also used as filter bed. The waste to be treated is diluted and distributed uniformly over the bed by nozzles. In order to remove treated liquid form the base of trickling filter and to provide adequate air supply, underdrains are provided.

Activated sludge method of biological treatment is similar to the trickling filter method in the sense that they bring liquid waste and air in contact for aerobic biological treatment. In activated sludge plant, liquid handled is in bulk and replenishment of dissolved oxygen is accomplished by bubbling air through liquid. The micro-organisms for flacculant growth are freely suspended in liquid instead of being attached to the rock or any other medium as in trickling filter. This biologically active growth is called **'activated sludge'**. The treatment is rapid. However, for its successful operation it is essential that adequate air supply is maintained throughout the liquid depth for sufficient aeration. Simplified flow diagram of activated sludge plant is shown below:

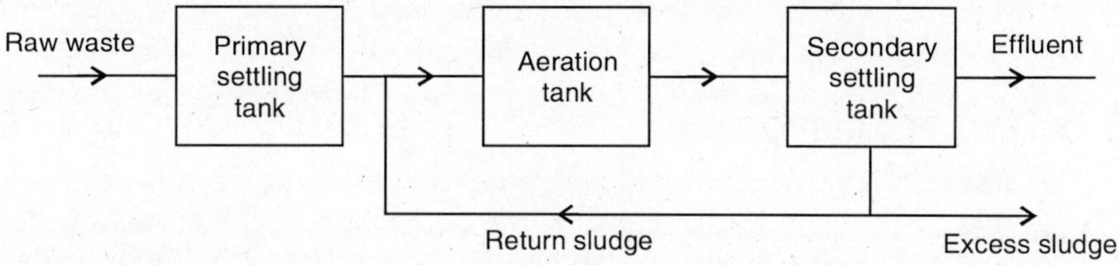

Fig. 13.5.

Lagooning or ponding in many locations is convenient and inexpensive method of treating industrial waste liquids. In the lagooning operation wastes are stored in shallow basins for period ranging from one day to several months. It requires considerable surface area. However, due to public safety, public nuisance, accidental release, ground water pollution etc. these are not suitable in urban and populated area.

(ii) Anaerobic Biological Treatment

Under anaerobic conditions *i.e.* in absence of oxygen breakdown of organic material is less oxygenated, for example, carbon items usually form methane (CH_4) instead of CO_2, Nitrogen becomes NH_3 or Ammonium compound or nitrogen gas instead of nitrites or nitrates. Sulphur compounds tend to form H_2S rather than SO_2.

Anaerobic fermentation and digestion is carried out by bacteria and similar micro-organisms. Anaerobic decomposition is characterised by ill smelling and noxious gases such as Ammonia and Hydrogen sulphide and by combustible gases such as methane, hydrogen. A major advantage of this type of treatment is that reactions can be carried out in depth, without need of large surfaces. However, final effluent is less satis-factory.

Many organic wastes when allowed to stand for several days become sour and acidic and produce methane and ill smelling gases. After several days of such activity the reactions gradually subside. Most of the organic matter is converted into CO_2, methane, water and stable solids.

Sludge obtained from sedimentation of industrial waste contains high concentration of organic matter, which can be subjected to anaerobic treatment called sludge digestion. The process is extensively employed in sewage treatment plants. The benefits of sludge digestion are enumerated as under :

1. Production of inoffensive sludge for disposal.
2. Reduction of quantity of sludge to be handled for disposal.
3. Production of sludge that will dry quickly.
4. Production of sewage sludge gas used as fuel.
5. Production of sludge for use as soil conditioner.

13.7. WASTE DISPOSAL

The industrial plant has to solve the waste problem either in the interest of good community relations or due to enforcement of Government rules and regulations. The waste may sometimes be discharged without further treatment in streams or municipal severage if permissible, considering the concentration of waste and dilution available in receiving stream. Otherwise, the industrial waste has to be treated before its disposal.

The method of waste disposal depend on the nature of waste. Most solid wastes are dumped on land as soil heaps or as landfill to quarries or mine shafts or as dumps consisting of a wide range of materials. Besides these some wastes are dumped into the seas. The various methods of waste disposal are illustrated in Fig. 13.6.

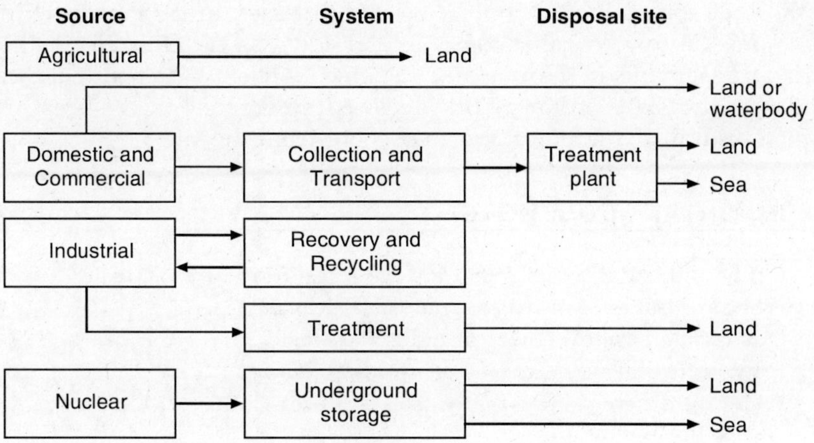

Fig. 13.6. Modes of waste Disposal

(i) Detoxification of Toxic Wastes

The toxic wastes are converted into less hazardous products by treatment with acids; cyanide containing wastes are decomposed by interaction with oxygen to form carbon dioxide and nitrogen. The decomposition can also be carried out biologically by means of suitable micro-organisms (bacteria). Sludge from petroleum refineries may be spread on the soil and left to decay into harmless products.

Incineration of toxic wastes is another method for their disposal. However, incinerators need pollution control devices and careful monitoring to make sure that they do not release toxic byproducts into the environment.

(ii) Land and Ocean Disposal

Radioactive wastes from nuclear power stations are generally fused in glass containers and lowered to the ocean floor. In USA such wastes are sealed in metal drums and buried underground at great depths. But they may leak or be damaged by earthquake and release the wastes into ground water.

Hazardous wastes dumped into soil/ditch have chances of leaking to the ground. A typical case history is that of the Love canal in Niagara Falls, New York, USA. In 1930 - 53 the canal ditch was the dumpsite for hazardous chemical wastes and municipal wastes. In 1953 the ditch was filled up; it was covered with clay and sold to the city Board of Education which built an elementary school. Some houses were also built. Soon the residents of these houses (300 families) and the school authorities complained of foul odour and illnesses. In 1978 it was found that some 25 toxic organic compounds known as car-cinogens, leaked into the basements in the area and dispersed into air. As a result of these findings, the state of New York declared emergency in the area and transferred all the families and the school from the site.

IMPORTANT QUESTIONS

1. Define industrial waste. What are the types of industrial waste?
2. What are different sources of industrial waste?
3. Discuss the harmful effects of industrial waste.
4. How industrial waste can be controlled?
5. What are different industrial wastes a process industry? Give its reason and rectification.
6. Classify the various treatment of industrial waste.
7. What is biological treatment of industrial waste?
8. Explain waste disposal.

CHAPTER 14

ENVIRONMENT STANDARDS

14.1. INTRODUCTION

ISO

ISO is a network of the national standards institutes of 148 countries, on the basis of one member per country, with a Central Secretariat in Geneva, Switzerland, that coordinates the system.

ISO is a non-governmental organization. Its members are not, as is the case in the United Nations system, delegations of national governments. Nevertheless, ISO occupies a special position between the public and private sectors. This is because, on the one hand, many of its member institutes are part of the governmental structure of their countries or are mandated by their government. On the other hand, other members partnerships of industry associations.

Therefore, ISO is able to act as a bridging organization in which a consensus can be reached on solutions that meet both the requirements of business and the broader needs of society, such as the needs of stockholder groups like consumers and users.

ISO Means

Since "International Organization for Standardization" would have different abbreviations in different languages ("IOS" in English, "OIN" in French for *Organization international de normalization*), it was decided at the outset to use a word derived from the Greek isos, meaning "equal". Therefore, whatever the country, whatever the language, the short form of the organization's name is always ISO.

Fig. 14.1.

How it all Started

International standardization began in the electro technical field; the International Electrotechnical Commission (IEC) was established in 1906.

Pioneering work in other fields was carried out by the International Federation of the National Standardizing. Associations (ISA), which was set up in 1926. The emphasis within ISA was laid heavily on mechanical engineering. ISA's activities came to an end in 1942.

In 1946, delegates from 25 countries met in London and decided to create a new international organization, of which the object would be "to facilitate the international coordination and unification of industrial standards". The new organization, ISO, officially began operations on 23 February 1947

How ISO standards benefit society

For businesses, the widespread adoption of International Standards means that suppliers can base the development of their products and services on specifications that have wide acceptance in their sectors. This, in turn, means that businesses using International Standards are increasingly free to compete on many more markets around the world.

For customers: The worldwide compatibility of technology which is achieved when products and services are based on international Standards brings them an increasingly wide choice of offers, and they also benefit from the effects of competition among suppliers.

For governments: International Standards provide the technological and scientific bases underpinning health, safety and environmental legislation.

For trade officials: Negotiating the emergence of regional and global markets, International Standards create "a level playing field" for all competitors on those markets. The existence of divergent national or regional standards can create technical barriers to trade, even when there is political agreement to do away with restrictive import quotas and the like. International Standards are the technical means by which political trade agreements can be put into practice.

For developing countries: International Standards that represent an international consensus on the state of the art constitute an important source of technological know how. By defining the characteristics that products and services will be expected to meet on export markets, International Standards give developing countries a basis for making the right decisions when investing their scarce resources and thus avoid squandering them.

For consumers: Conformity of products and services to International Standards provides assurance about their quality, safety and reliability.

For everyone, International Standards can contribute to the quality of life in general by ensuring that the transport, machinery and tools we use are safe.

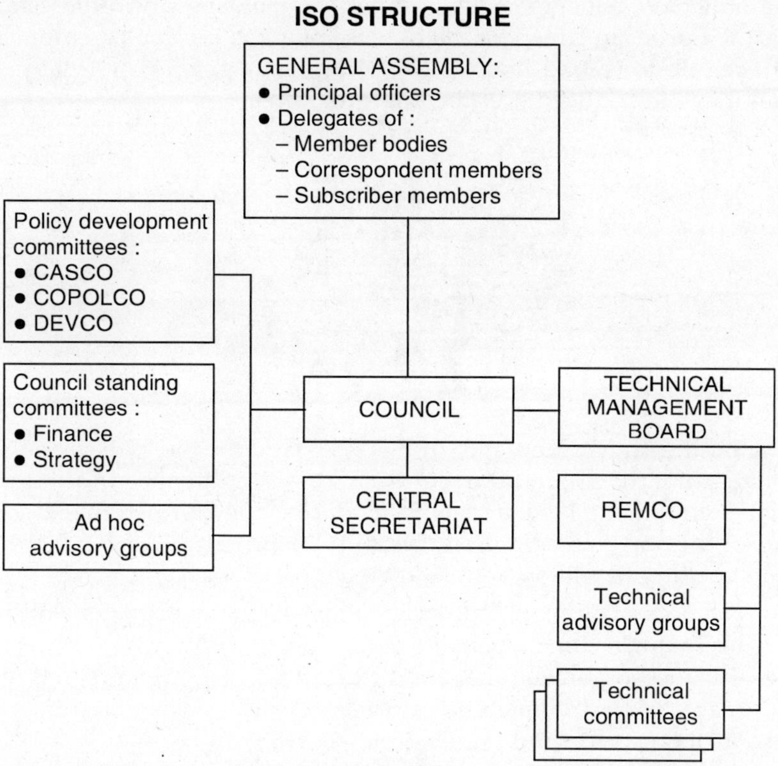

Fig. 14.2.

14.2. ISO14000 - INTRODUCTION

After the sucess of the ISO 9000 series of quality standards, the International Standards Organization is nearing completion and publication of a comprehensive set of standards for environmental management.

This series of standards is designed to cover the whole area environmental issues for organizations in the global marketplace.

↳ History of Development

The ISO 14000 series emerged primarily as a result of the Uruguay round of the GATT negotiations and the Rio Summit on the Environment held in 1992. While GATT concentrates on the need to reduce non-tariff barriers to trade, the Rio Summit generated a commitment to protection of the environment across the world. The environmental field has seen a steady growth of national and regional standards. The British Standards Institution has **BS 7750**, the Canadian Standards Association has environmental management, auditing, eco-labelling and other standards, the European Union has all of these plus the **eco-management and audit regulations**, and many other countries (e.g. USA, Germany and Japan) have introduced eco-libelling programs.

After the rapid acceptance of ISO 9000, and the increase of environmental standards around the world, ISO assessed the need for international environmental management standards. They formed the Strategic Advisory Group on the Environment (SAGE) in 1991, to consider whether such standards could serve to:

- Promote a common approach to environmental management similar to quality management;
- Enhance organizations' ability to attain and measure improvements in environmental performance; and
- Facilitate trade and remove trade barriers.

In 1992, SAGE's recommendations created a new committee, TC 207, for international environmental management standards. The committee, and its sub-committees include representatives from industry, standards organizations, government and environmental organizations from many countries. The new series of ISO 14000 standards are designed to cover:

– *environmental management systems*
– *environmental auditing*
– *environmental performance evaluation*
– *environmental labelling*
– *life-cycle assessment*
– *environmental aspects in product standards*

↬ What does the ISO 14000 Series cover?

Table 14.2.

Standard	Title/Description
14000	Guide to Environmental Management Principles, Systems and Supporting Techniques
14001	Environmental Management Systems - Specification with Guidance for Use
14010	Guidelines for Environmental Auditing - General Principles of Environmental Auditing
14011	Guidelines for Environmental Auditing - Audit Procedures - Part 1: Auditing of Environmental Management Systems
14012	Guidelines for Environmental Auditing - Qualification Criteria for Environmental Auditors
14013/15	Guidelines for Environmental Auditing - Audit Programmes, Reviews & Assessments
14020/23	Environmental Labeling
14024	Environmental Labeling - Practitioner Programs-Guiding Principles, Practices and Certification Procedures of Multiple Criteria Programs
14031/32	Guidelines on Environmental Performance Evaluation
14040/43	Life Cycle Assessment General Principles and Practices
14050	Glossary
14060	Guide for the Inclusion of Environmental Aspects in Product Standards

14.3. GENERAL DESCRIPTION OF ISO 14001

ISO 14001 requires an Environmental Policy to be in existence within the organization, fully supported by senior management, and outlining the policies of the company, not only to the staff but to the public. The policy needs to clarify compliance with Environmental Legislation that may effect the organization and stress a commitment to continuous improvement. Emphasis has been placed on policy as this provides the direction for the remainder of the Management System.

The Environment Policy is different; this provides the initial foundation and direction for the Management System and will be more stringently reviewed than a similar ISO 9000 policy. The statement must be publicized in non-technical language so that it can be understood by the majority of readers.

The preparatory review and definition of the organization's environmental effects is not part of a ISO 14001 Assessment, however examination of this data will provide an external audit with a wealth of information on the methods adopted by the company. The preparatory review itself should be comprehensive in consideration of input processes and output at the site. This review should be designed to identify all relevant environmental aspects that may arise from existence on the site. These may relate to current operations, they may relate to future, perhaps even unplanned future activities, and they will certainly relate to the activities performed on site in the past (i.e. contamination of land).

The company will declare its primary environmental objectives, Those that can have most environmental impact. In order to gain most benefit these will become the primary areas of consideration within the improvement process, and the company's environmental program. The program will be the plan to achieve specific goals or targets along the route to a specific goal and describe the means to reach those objectives such that they are real and achievable. The Environmental Management System provides further detail on the environmental program. The EMS establishes procedures, work instructions and controls to ensure that implementation of the policy and achievement of the targets can become a reality. Communication is a vital factor, enabling people in the organization to be aware of their responsibilities, aware of the objectives of the scheme, and able to contribute to its success.

As with ISO 9000 the Environmental Management system requires a planned comprehensive periodic audit of the Environmental Management System to ensure that it is effective in operation, is meeting specified goals, and the system continues to perform in accordance with relevant regulations and standards. The audits are designed to provide additional information in order to exercise effective management of the system, providing information on practices which differ to the current procedures of offer an opportunity for improvement.

14.4. ENVIRONMENTAL MANAGEMENT SYSTEM (EMS)

An environment management (EMS) is a systematic approach to dealing with the environmental aspects of an organization. It is a 'tool' that enables an organization of any size or type to control the impact of its activities, products or services on the natural environment. The ISO 14001 standard *"Environmental management systems. Specification with guidance for use"* is the standard within ISO 14000 series that specifies the requirements of an environmental management system.

↬ Benefits of an EMS

- Assuring customers of commitment to demonstrable environmental management
- Maintaining good public/community relations
- Satisfying investor criteria and improving access to capital
- Obtaining insurance at reasonable cost
- Enhancing image and market share
- Meeting vendor certification criteria
- Improving cost control
- Reducing incidents that result in liability
- Demonstrating reasonable care
- Conserving input materials and energy
- Facilitating the attainment of permits and authorizations
- Fostering development and sharing environmental solutions
- Improving industry-government relations

14.5. KEY ELEMENTS OF ISO14001 EMS

↬ Environmental Policy

- the environmental policy and the requirements to pursue this policy via objectives, targets, and environmental programs

↬ Planning

- the analysis of the environmental aspects of the organization (including its processes, products and services as well as the goods and services used by the organization);

↬ Implementation and operation

- implementation and organization of processes to control and improve operational activities that are critical from an environmental perspective (including both products and services of an organization)

↬ Checking and corrective action

- checking and corrective action including the monitoring, measurement, and recording of the characteristics and activities that can have a significant impact on the environment

↬ Management Review

- review of the EMS by the orgainization's top management to ensure its continuing suitability, adequacy and effectiveness

Continual improvement

- the concept of continual improvement is a key component of the environmental management system; it completes the cyclical process of plan, implement, check, review and continually improve.

IMPORTANT QUESTIONS

1. What does ISO stands for?
2. What does ISO means and how it all started?
3. How ISO standards benefit society.
4. How ISO 14000 was introduced?
5. Give general description of ISO 140001.
6. Define 'EMS'.
7. What are the benefits of EMS.
8. What are the key elements of ISO 140001?
9. What are the various environment standards for representative industries.

CHAPTER 15

Occupational Ergonomics

15.1. INTRODUCTION

Below are a few examples of what people have sat down on over the years :

European folding-chair from 11-200 B.C. originally with a seat of otter skin.

Fig. 15.1.

Chair and "table" from the Middle Ages. About 1200

Fig. 15.2.

Beautifully carved wooden chair from 1100. Oslo University.

Fig. 15.3.

Vincent van Gogh's famous chair Arles, painted in 1888

Fig. 15.4.

Office chair from the sixties

Fig. 15.5.

Modern office chair from 1995 with separately adjustable seat and back.

Fig. 15.6.

Ergonomics is a multi-disciplinary science which emphasizes the importance of designing workstations (*i.e.* office furniture or industrial work areas and equipment) to fit the individual worker.

Ergonomics provides the scientific basis for optimized design of the work environment compatible with the capabilities and limitations of the working population. An ergonomic is trained to recognize, and control hazards in the work environment that result from a poor fit between the worker and the workplace; these hazards may result in acute injure, chronic musculoskeletal disorders, or mental/psychosocial stress. The physical demands of machines, tools, and work methods must accommodate the range in size, strength, mobility, and endurance of the workforce. Information flow must be structured so that the worker can process and respond appropriately without being mentally overstressed to the degree that errors result. Job content and work organization should be designed to optimize skill utilization and learning, physiological and psychological well-being.

The concentration in Ergonomics is designed to develop an understanding of human anatomy, physiology, and psychology, of industrial hygiene and epidemiology as well as modern manufacturing technology and work organization. It provides a multidisciplinary background in these basic areas, as well as their application to the practical problems in ergonomics that are encountered in industrial and service environments

Good ergonomics adapt the job to fit person rather than forcing the person to fit the job. In an ergonomic workplace tasks and tools are designed to fit individual capabilities and limitations so people can do their jobs without being injured. Ergonomics has emerged as a hot issue because it shows the link between certain types of injuries and the ways in which people perform their jobs. The study of ergonomics–both in general and in terms of specific tasks and motions–has helped to identify what types of positions and movements can cause physical pain and injury, as well as ways to prevent these problems.

15.2. THE HUMAN BODY

The Human Spine

The spine is literally the supporting structure of the body. In profile, the spine is curved in a faint S-shape to help it cope with the different kinds of load. It is this natural form that should be supported during sedentary work. Any modern office chair can do this but there are natural limits to it. We are not designed to still like that even in the correct position, for long periods of time, will create problems. Backaches will remind you when enough is enough.

The spine is well sprung with resilient discs between the vertebrae and luckily, too. Otherwise, just one ordinary step would result in a minor concussion. The sacrum and coccyx cannot move but the 24 other vertebrae are connected by moveable joints. Together they make a very elastic construction.

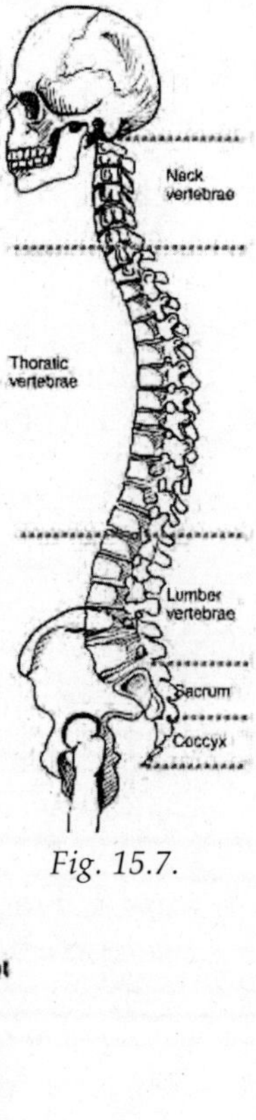

Fig. 15.7.

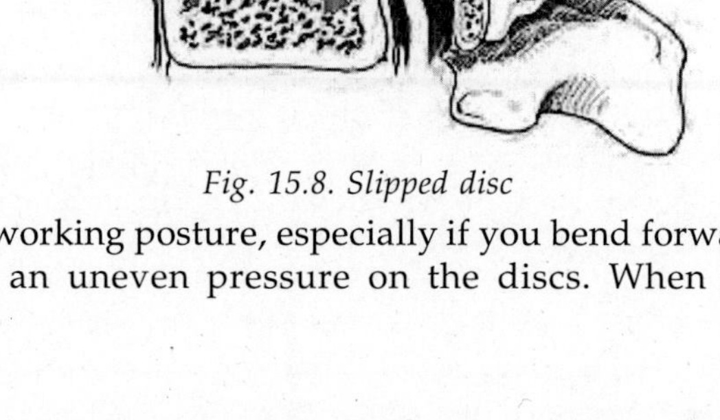

Fig. 15.8. Slipped disc

A wrong working posture, especially if you bend forward too much, will produce an uneven pressure on the discs. When the discs are "squeezed" out of shape, the surrounding tissue will compress or stretch. If the pressure becomes excessive, you may end up with a slipped disc. The outer ring of the disc will rupture so that the soft core will be squeezed out and press on the nerves.

Muscles

Muscles like to work. They almost never hurt when they are used but hurt only when they are abused.

Pain in neck and shoulders is usually caused by a wrong position with the neck bent forward and the shoulders raised.

When you move, a muscle will alternately tighten and relax. This action keeps blood circulating to remove toxins from the muscles and supply it with fresh oxygen and nutrients.

Problems will arise if a muscle is more or less tight all the time as when working with a wrong posture. The muscle is unable to get rid of toxins and receives no fresh oxygen. What you feel is pain, a warning signal, the body's way of protecting the muscles against overload.

Regular pauses are no a waste of time. They are a must for our well-being and even for our working efficiency. Too often many people tend to ignore the signals and persevere bravely to the point of risking severe muscle aches.

15.3. HEART

The **heart** is a remarkable device. It's a fully automatic pump capable of pumping up to 35 litres of blood a minute under special circumstances. It could fill an ordinary tub in less than 10 minutes.

Other parts of the circulatory system are not quite so automatic, they need a little help from the muscles, again in the shape of movement, for it is their job to pump the blood back to the heart by pressing on the veins.

The circulatory system is a closed system and must run non-stop without interruption. The vene pump, as it is called, needs help all the time when you are sitting still. It's important to use the leg muscles so that they are neither tight nor relaxed for any length of time.

15.4. POSTURE

Proper Seating and Sitting Properly

It all begins with the chair. In the past the chair has been expected to provide both sitting and working comfort for employees who work at fixed height surfaces. This height, twenty-nine inches for working desk height and twenty-six inches for typewriters, was determined years ago to suit the physical requirements of the 'average' individual.

The unfortunate result has been the necessity for shorter than average individuals to raise their chair as they try to reach a comfortable working position at a work surface which is too high for them.

Sitting back in the chair would then result in the legs either hanging with feet off the floor, or atleast, the individual may experience pressure under the front of the thigh where it presses against the chair at an angle.

As this is a sensitive part of the leg, through which pass many nerve endings and veins, the user if forced to sit forward in the chair to avoid discomfort.

In fact in any large office with fixed height desks, you will see an unacceptable number of shorter individuals, especially female, forced to sit forward on the chair, away from the support of the chair back. For all the benefit the chair back is giving them; they might as well be sitting on stools.

Employees who are taller than average are also affected by the fixed height work surfaces. Though they would more comfortably sit in a higher chair, the work surface would then be too low, causing them to slouch forward. So instead they lower their chairs, only to cause other problems.

When the individual sits back in his lowered chair, his knees will be higher than his behind when his feet are properly flat on the floor, causing all his weight to be on the back of the chair.

Every time this individual sits back in his chair, the chair will tilt. Then every time he wishes to work at his desk, he must sit forward in the chair away from the support for his back.

As with the shorter than average individual, this has to have some effect. Though the degree of back pain, fatigue or aggravated tension experienced will vary, it will cause discomfort. In fact, backache is the third most widespread illness in our population.

Fig. 15.9.

The proper method of adjusting a chair for sitting comfort is to stand facing the chair and adjust it to a height just below the kneecaps.

Then when you sit back in the chair your thighs will be parallel to the floor and your weight evenly distributed across the chair. While there are a regrettable number of chairs still being made with high and prominent arms that prevent the user from sitting properly positioned at his or her desk, the trend is towards chairs with recessed or sloping arms that do not cause this obstruction. Such chairs, of course, are highly recommended.

Finding The Right Position

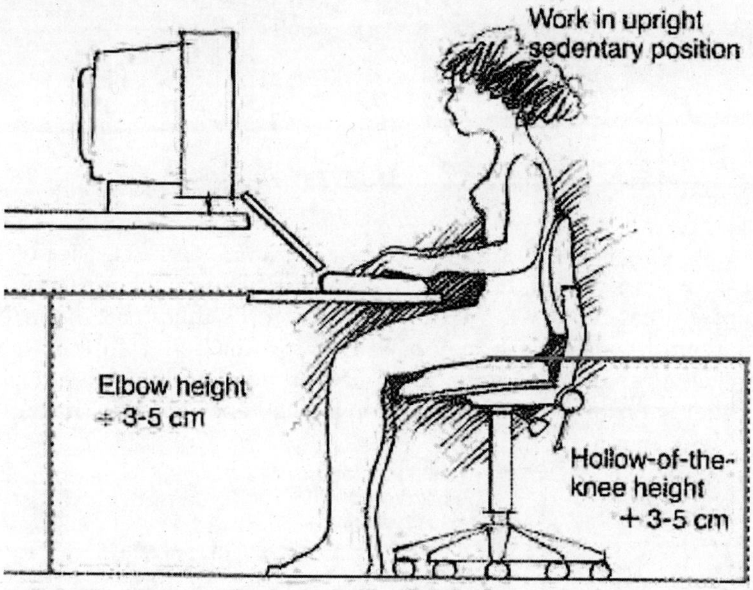

Fig. 15.10.

For screen work you may use the classic position with everything at right angles: neck thigh, elbows and legs, and with your feet planted solidly on the floor. The chair must be adjusted so that it will support the natural S-shape of your spine.

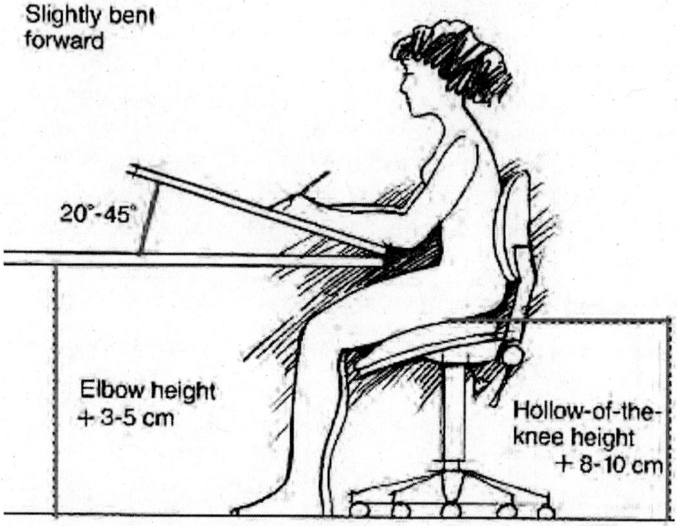

Fig. 15.11.

The reading and writing positions is a variation of the classic position. It's best for your back if the seat of your chair can follow your movement forward so that there will be no pressure on the thigs and the arteries under them. A slanting tabletop will both offer better support and a better visual angle. Use your imagination-move about on your chair. Find other ways of sitting. No position is good enough to be the one and only.

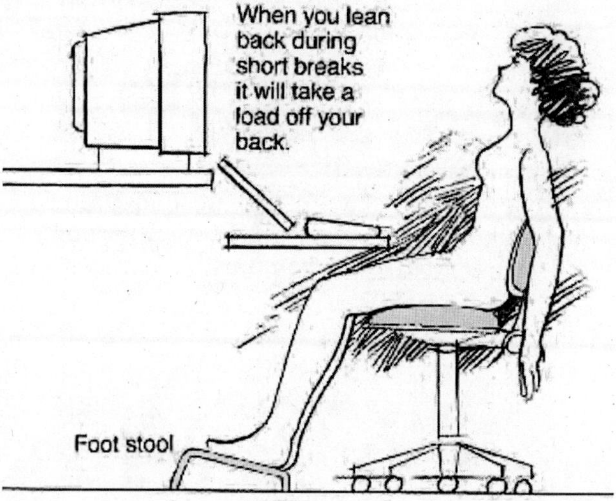

Fig. 15.12.

Both back and seat can tilt on modern chairs. It's good for the back if you sometimes lean back and rest in this position-but not for too long on account of your neck muscles.

Requirements of Working Height

Once employees adjust their chairs to the comfort requirement of their heights, they then will be seated at differing heights relative to their work surfaces, typewriters or terminal keyboards.

If these keys or surfaces are too high, then the users are forced to raise their arms. Many shorter individuals raise their arms from the shoulders, almost like having a constant "shrug". This uses muscles across the top of the shoulders and at the base of the neck, and it often takes very little time before discomfort is felt.

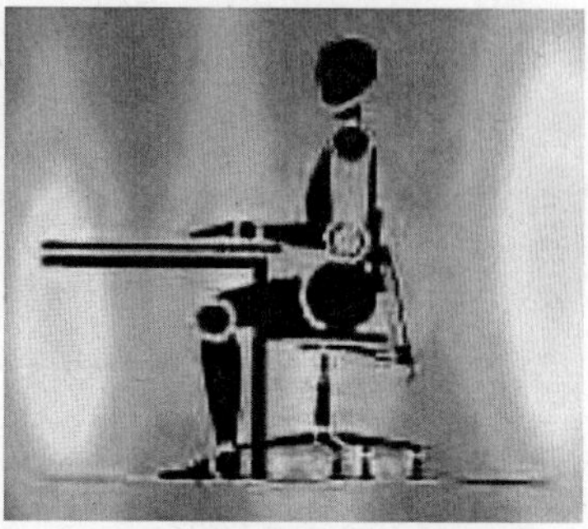

Fig. 15.13.

At a properly positioned work surface, terminal keyboard, or typewriter, the height of the surface or keys will be at elbow height, when the individual drops his arms to his side while sitting. At this height the employee may work in an upright, yet comfortable sitting posture, with the forearms at right angles and parallel to the floor in a relaxed working position.

When considering that the average values shown above are for males and considering again that the values for the female population would cover a similar range, but due to the fact that females generally are physically smaller it suggests itself to question the wisdom of a standard of 29.5" high work surface (desk top) which is the current standard height in North America when 29.7" is considered maximum height for a man.

15.5. WORKPLACE DESIGN OR OFFICE ERGONOMICS

In the past, many workplaces were being designed (and unfortunately still are) using stale formulas rather than a process involving the study of strategic needs and cost-effective solutions that are appropriate for an individual company.

In the rapidly growing/changing business model of the past few years the focus has been on the ability to rapidly move into a space and start work instantly, also stressing the ability to expand quickly–immediate fulfilment of needs.

Fig. 15.14.

Using large open spaces and a "work anywhere" concept, the workplace was designed to be a dramatic, sensory experience with total flexibility and an emphasis on teaming.

Today's workplace design is being affected by the changing economy. There is more value placed on the philosophy that there is a direct relationship between the success of a business and the design of the work environment. There are quite a few benefits that a company should expect from a well designed office environment. Good design in the workplace can elevate morale and improve productivity, promote the image and help reinforce a corporate identity. It can help to attract and retain a higher quality workforce.

The workplace design depend upon many factors like

Office Culture

The starting point for any design team must be to have an understanding of it's client's culture and business philosophy. Asking the right questions is essential: Why is this company in business? What are the values of this company? What is the vision? Who are the customers or clients? Who are the people that work here?

Corporate Identity

What is the visual image and message that is conveyed to a customer, client or employee? What is the mission of the company? This is a critical element to either help a business that is starting out to grow or help a long-standing business progress into the 21th century. The carefully planned design and ambiance that is created will support and promote the identity of the company.

Organizational Structure

Understanding the nature and the process of the workflow in a company is essential in developing a good design program. How does the work come in the door? How does information get processed, communicated and transmitted from one place to another? How can the design of the workplace enhance the efficiency of the company's workflow?

Technology

Technology is rapidly changing the way companies work, but in the midst of the cell phones and e-mail, face-to-face meetings are string critical . With the probability of work, people can collaborate in a variety of ways that are not always **in person**. Therefore, physical accommodations must be planned for within the work environment to support new technology and its interface.

Flexibility

It would not be possible to overstate the importance of flexibility in the design of today's workplace. The speed at which technology and markets change have created the need for companies to grow, shrink or change focus altogether. What are the company's goals for staffing growth? In two years? In five years? Will there be a need to accommodate telecommunicating, flex-time and job sharing? Are there opportunities for flexible space when it comes to conferencing and multimedia or lounge and informal open meeting areas?

Good design should help a company realize its goals. A well-designed work environment will meet the business requirements of a company, maximize the productivity of employees, attract and retain a high quality workforce, increase profits, foster repeat clients, and support building new business.

15.6. ERGONOMICS FOR WOMEN AT WORK

- Women currently comprise 46% of the 137 million workers in the United States, with their share of the labour force projected to reach 48% by 2008.
- In 1999, 75% (46 million) of employed women worked full-time, while 25% (16 millon) worked part-time.
- In 1999, 3.7 millon women held multiple jobs
- Sixty Percent of women age 16 and over were either employed or looking for work in 1999.

- Of employed women, 40% held technical, sales, and administrative support positions; 32% worked in managerial and professional specialities; and 17% worked in service occupations in 1999.

Men and women are not the same and the jobs they do, their working conditions and how they are treated by society are not the same. These factors can affect the hazards they face at work and the approach that needs to be taken to assess and control them. Factors to take into account include:

- *women and men are concentrated in certain jobs, and therefore face hazards particular to those jobs*
- *Women and men are physically different, including in regard to reproduction*
- *Women and men have different responsibilities in the home. Working women may have two jobs, **at work and in the home** .*

It is therefore important to recognize these differences and take a "gender sensitive" approach to health and safety at work. Women have fewer accidents than men, but more industrial disease. The official statistics may underestimate women's health problems, due to discrimination against women in the compensation system, inadequate ways of treating part employment and delay in identifying industrial disease in women's jobs.

Musculo problems are the women's chronic health problem and can be related to the static muscular effort and highly repetitive work found in many jobs assigned to women

Stress is reported by a very large number of women workers. Emotional and mental stresses have multiple causes and effects, and do not fit neatly into the standard occupational health practices based on single limited effect phenomena like asbestosis. The increasing diversity of schedules in services and industry makes it very difficult to maintain a healthy family life, still the responsibility primarily of women.

Attention should be paid to the working conditions of cleaners, hairdressers, waitresses and other service workers. Methods should be developed for diminishing exposure to chemicals among hairdressers and laboratory technicians. Attention to women's reproductive health should not be limited to fetal protection but should also include attention to fertility, sexuality, early menopause and menstrual disorders. It is important that pregnant and nursing women be given rapid access to up information on occupational and environmental risks which may affect them, as well as to programs for protection of their health. Occupation related dysmenorrhoeal should be recognized and measures taken to diminish incidence and to provide for compensation of related absences. Methods should be developed in the workplace for protecting the psychological health of workers who are forced to buffer institutional requirements and the needs of clients. Finally, gender sensitive and women occupational health research should be encouraged and supported. It should be recognized that such research may require innovative methods, and must involve the active participation of working

15.7. PHYSICAL WORK AND ENVIRONMENT

Employees' exposure to physical working environment factors

The share of people in work/employment who report exposure to various factors in the working environment varies widely among women and men, blue-collar and white-collar workers, and - not least - employees in different occupations. Agriculture, for example, provides one of the most dangerous working environments.

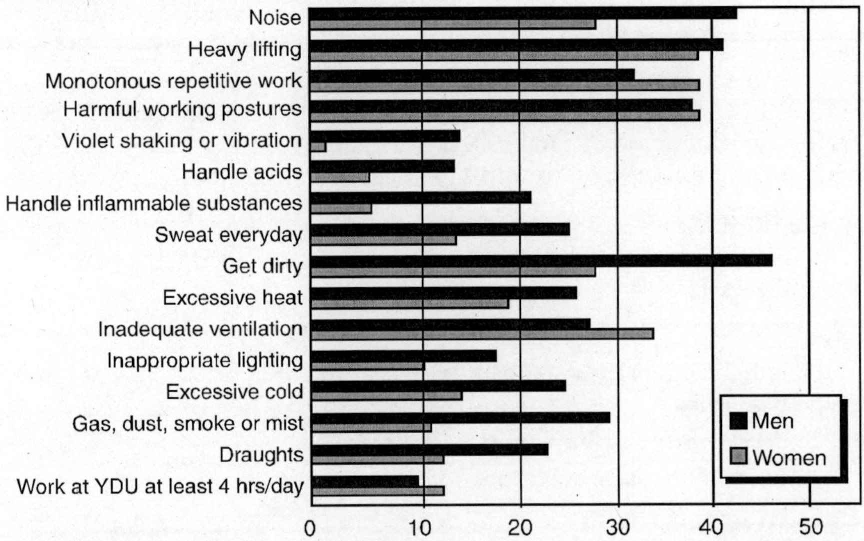

Fig. 15.15.

Problems in the physical working environment are most widespread in blue-collar occupations but some of them do also occur in some white-collar occupations.

The physical characteristics of the working environment should ideally be of such quality that no offence is given to any of the five senses.

Some of the physical characteristics of working environment are:

1. Noise

Noise can be considered form the point of view of its

- intensity,
- frequency spectrum,
- steady state or transient character,
- influence on human well-being and/or working efficiency,
- influence on health.

All these factors are to a certain extent interdependent, which must be taken into account when evaluating a given situation

2. Vibration

Vibrations may emanate either from fixed installations, such as large compressors, machines or machine-tools, bubble chambers (in the past) or from hand-held equipment, such as pneumatic hammers, drills, etc. Road traffic is another source of vibrations not to be forgotten.

If the source is hand-held, the transmission of the vibrations is of course direct to the hands and arms. In the case of a fixed source or road traffic, vibrations are transmitted via walls, floors and foundations, occasionally following odd and unexpected trajectories.

3. Lighting

An adequate lighting situation can only be obtained by considering several factors

Intensity

The light intensity must be adapted to the area or the task at hand. Some examples:

Exterior traffic or movements-20 lux

Interior staircases, corridors-150 lux

Contrasts

For optimal lighting, contrasts should be limited, preferably to a radio not exceeding 1:10 and back-lighting of the work piece should be avoided. Glaring light sources in the visible surroundings must be properly masked to keep the contrasts at a comfortable, non-stressing level.

Reflections

Reflections, whether on the work piece, on supporting or on surrounding surfaces should be kept as low as possible by adjusting the lights, the orientation of the workstation or by the use of non-reflecting materials where appropriate. Screens may also be of help.

Quality of Light

The colour characteristics of the light source should be adequate for the work at hand. The choice is normally between incandescent lamps (warm light, smooth colour spectrum) and fluorescent lamps (medium to cool light, colour spectrum made up of more or less dense peaks). Where a good colour rendition is important this must be taken into consideration.

Fluorescent lamps may sometimes have a slight flicker, not immediately noticable but tiring and irritating in the long run.

Maintenance

Lamps of all types age, slowly and unnoticeably. Not only does the light output decrease, but the quality of the light deteriorates, flickering and unsteadiness increases and the colour characteristics changes. To keep this deterioration of the light quality within tolerable limits, lamps should be replaced, in the framework of the regular maintenance service, well before their actual death.

4. Air Quality

Amongst the factors that influence the thermal well-being are:

- the physical effort,
- the clothing worn,
- the humidity of the air,
- the temperature of radiating sources (sun, lamps, ovens, etc.),
- draught, the speed of the air (ventilation system, open widows, etc.).

5. Work Space

Some factors to consider are:

- inter-personal relationships,
- crowding,
- conflicts between smokers and non-smokers,
- the interest that the management shows in the work-situation of the individual

15.8. ANTHROPOMETRY

Anthropometrics: *the measurement of the physical characteristics of humans, particularly length and weight measures, used for industrial design, especially the design of furniture, vehicles, and workspace.*

In the domain of computer use, such measurements apply primarily in the design of input and output devices, computer desks and chairs, and in the application of recommended working postures. The typical goal is to understand not just the average measurements of, for instance, arm length and joint rotation angles, but to understand the range of possible variation within the target user population to adequately support the full range of users, from children to aging adults, and including both male and female dimensions.

Anthropometry is also commonly applied by forensic anthropologists in order to correlate unknown human remains with missing persons. Measurements of remain are compared to statistical models based on known skeletal collections to determine probability for various characteristics about unknown remains, such as stature, age at death, sex, and race.

Anthropometric data is used in many areas of manufacture to provide information for the design of products such as clothing, footwear, safety equipment, furniture, vehicles and any other objects with which people interact. In the armed services this data is of particular importance as the survival of the serviceman may depend upon it. For example, in the accurate shape of close fitting protection equipment such as body armour or combat helmets, or in the optimum ergonomic design of combat vehicles. Also, accurate anthropometric data is required in the determination of size ranges and stocks for both clothing and personal equipment, and the increased use of new materials and designs in military clothing will place additional demands on the accuracy of fit for future clothing systems. Elsewhere, the increasing use of the Internet for on-line electronic retailing has opened up exciting new uses for anthropometric data in virtual clothes shopping applications.

IMPORTANT QUESTIONS

1. What is ergonomics?
2. Discuss how ergonomics is related to working environment?
3. Discuss the importance of 'Postures' in working.
4. Explain the importance of ergonomics for women at work.
5. What is 'Anthropometry'?

□□□

CHAPTER 16

OCCUPATIONAL STRESS

16.1. WORK-RELATED STRESS

Work-related stress is the natural reaction of people to being put under intense pressure at work over a period of time. Many people are motivated by the challenges and difficulties that normally occur with work demand and react by improving performance. Meeting those challenges and overcoming the difficulties causes feelings of relaxation and satisfaction. When the pressure of work demands become excessive and prolonged, however, people perceive a threat to their well-being or interests and then experience unpleasant emotions such as fear, anger or anxiety.

The basis of this response comes from instinctive "fight or flight" reactions to danger which has evolved from early history when people live in far more threatening natural environments.

The stress response is designed to be used in short bursts and then switched off. If it is achieved for too long, or the period between stressful situations is too short, then the body has no time to repair itself, and fatigue and damage occurs. The stress hormones then literally begin to destroy the body so, over time, this affects physical and mental health and quality of life in just the same way as exposure to industrial toxins.

16.2. CAUSES OF STRESS

It is reasonable to assume that people in the workplace are fit to cope with the pressure from normal work demands.

The reactions of individuals will vary according to the nature of the pressures and the extent to which the person is directly involved. What may be seen as a challenge by one person may be an impossible task or boring and repetitious to another.

A person's background, motivation, experience, skills and knowledge on the one hand and the support and encouragement from managers, supervisors and colleagues on the other, all play an important role.

While it may be beyond the employer's or supervisor's responsibility or expertise to overcome, it is necessary to take into consideration that people will come to work with various character traits and in various moods.

Problems outside of work can also contribute to stress *e.g.*, relationship problems and financial pressures. Non-work problems can make it difficult for cope with the pressures of work, and their performance at work may suffer. A death or sickness in the family, a temporary setback or other personal problems may exacerbate the situation and influence the way people cope with work pressures.

The causes of stress (often called stressors) can be many and varied, and can occur as a result of combinations of more than one stressor. Most stressors can be grouped into one of three areas. For each group of stressors are suggestions for managers and supervisors to consider when dealing with those triggers.

Stress from doing the job . . .

- Boring or repetitive work, or too little to do.
- Too much to do, too little time.
- To little/too much training or support for the job.
- Confusion over, or too much, responsibility for managing others.
- Confusion over priorities, time frames and standards.
- Sub-standard performance resulting in disciplinary action such as dismissal, retrenchment, demotion, discipline, transfer or redeployment, a worker not being promoted, reclassified, transferred or granted leave of absence or any other benefit in relation to employment or a worker's expectations of any of these matters or of the employer's decision in relation to these matters.

. . . and what to do about it

- Change the way jobs are done, consult with individuals before moving them to comparable work, given individual more responsibility, increase job scope, vary job tasks, give workers responsibility for group performance.
- Try to give warning of urgent or important tasks, priorities tasks, remove unnecessary work.
- Match individuals to jobs, increase the scope of jobs for the over-trained.
- Ensure everyone has clearly defined and achievable objectives and responsibilities linked to organizational objectives.
- Provide supervisor training and support.

- Ensure disciplinary procedures are developed and applied by trained and authorised officers able to communicates effectively with employees.
- Give staff the knowledge and skills to enable them to mange their own stress response.
- Ensure staff and supervisors receive the skills, training and resources they need to work purposefully, confidently and are appreciated.
- Ensure that policies and procedures are clearly stated and uniformly applied.

Stress from Work Relationships . . .

- Poor relationships with colleagues.
- Bullying or harassment.
- Lack of communication or consultation between manager and employee.
- Negative culture based on blame for and denial of problems.
- Misguided practical jokes or initiation ceremonies.

. . . and what to do about it

- Provide training in interpersonal skills.
- Establish and review policies, grievance procedures and investigation of complaints.
- Provide opportunities for employees to contribute ideas and to be consulted.
- Encourage management lead by example with honesty, respect and support.
- Ensure people are treated fairly and consistently with an intolerance of harassment and bullying.
- Provide opportunities for social interaction among workers.

Stress from Working conditions . . .

- Physical danger, or the threat of it.
- In contact with human suffering and people's reactions to it.
- Single incident such as armed robbery or a workplace fatality.
- Poor physical working conditions.
- Inflexible work schedules or unpredictable hours.
- Organizational change, such as restructing.

. . . and what to do about it

- Provide adequate control measures against physical pressures, in particular, clear support for those faced with violent situations.

- Consultation with staff over restructuring plans.
- Encourage managers to have an open and understanding attitude towards those who admit to being under too much pressure, especially if employees are fearful of losing their jobs.
- Provide scope for flexible and variable working conditions over which people have some control, to increase interest and ownership of the job, and assist people to meet their family responsibilities.
- Engender consultation and cooperation between employers, employees and their representatives, especially during periods of organizational change.
- Evaluate the effectiveness of changes put in place and review the situation form time to time.

There are also some general strategies relevant to most organisations that may help:

- Encourage employers to take stress seriously and make themselves aware of the warning signs of people under too much pressure.
- Encourage staff to talk about the pressures they perceive in their jobs;
- Be understanding of stressors outside the workplace, but only seek information you need to know.
- propose and prioritise intervention strategies and inform employers; and
- Encourage employers to provide a consistent management approach.

Larger employers may have the resources to provide stress management training, a confidential employee assistance programme or use risk management consultants as tools to assist with stress problems in their organization. These tools may be useful as part of a wider plan to tackle organizational issues but they are not always effective in identifying the source of a stress problem. Smaller employers may need to consider more affordable options.

16.3. SIGNS OF STRESS

Personal Signs of Stress

- immediate body changes that may be associated with distress, such as changes in heart rate and breathing rate, muscle tension, nausea or vomiting;
- feelings of anger, protest or frustration;
- feelings of anxiety or being out of control;
- feelings of guilt or embarrassment;

- irritability and loss of concentration, poor memory, learning difficulties;
- prolonged sleeplessness and disturbing dreams.

Outward signs of stress in individuals that could be noticeable to colleagues and managers:

- deteriorating relationships with colleagues;
- irritability, indecisiveness, absenteeism, reduced performance;
- demand for more precise instructions;
- increased smoking or alcohol consumption, or drug taking;
- complaining about ill-health;
- actual ill-health such as frequent headaches, gastro-intestinal disturbance, ongoing minor illness, skin rashes, deteriorating immune response.

Signs of stress across an organization that represent a businesscost:

- high staff turnover, increased absenteeism, reduced work performance, poor timekeeping and more customer complaints;
- stress appearing in those people who cover for the stressed colleague who is absent from work;
- increased rate of workplace accidents;
- compensation claims from people whose unchecked stress results in ill-health;
- reduction in output, product quality, service or morale;
- increased wages/overtime payments;
- deterioration of industrial relations;
- organizational sabotage.

16.4. DUTIES OF EMPLOYERS AND EMPLOYEES

Stress becomes an occupational hazard if it adversely impacts on safety and health in the workplace.

Often a number of factors from the workplace and home combine to increase stress to the point where a person may put their own or another person's safety at risk. As a result, both employers and employees have a special duty of care role to play in making sure risks that are caused by stress from all sources are minimised and staff are fit for work.

Employers have a duty to provide a workplace where, so far as practicable, employees are protected from hazards in their workplace. This includes providing safe systems of work, information, instruction, training and supervision. There is also a requirement for employers to consult and cooperate with safety and health representatives and other employees.

Employees must take reasonable care for their own safety and health and avoid adversely affecting the safety and health of others in the workplace. They must also cooperate with employers in safety and health matters so that employers can act responsibly.

The *Occupational Safety and Health Regulations 1996* require employers, where practicable, to adopt a systematic approach to identifying, assessing and controlling hazards at work. Employers should identify factors in the workplace that cause stress (a number of which are have already been listed). They should then assess the likelihood of stress causing injury or disease. If the risks are significant, they should put in place controls to minimise them.

16.5 MEASUREMENT OF STRESS

It is difficult to objectively measure the level of stress in the workplace. For example, there are no simple tests to measure how much pressure a person is under and then assess their ability to work safely.

The design and appropriate solutions identified from stress prevention programmes will vary according to the size and complexity of the organization, available resources and especially the unique types of stress problems faced by the organization.

In spite of difficulties in measuring stress, both employers and employees regularly make judgements about levels of stress and the level of associated risk. The absence of objective tests does not remove the need to do a risk assessment nor does it mean that the assessment is invalid.

Responsibility for risk assessment rests with the employer, main contractor, self-employed person or person with control over the workplace or access to it, and involves:

- identifying pressures at work which could cause high and long-lasting levels of stress;
- identifying people whose might be affected by these pressures;
- deciding whether preventative action in place is sufficient;
- taking action to prevent pressure growing to an unacceptable level; and
- reviewing the success of the control measures.

It is important that there is early intervention if stress is identified. Risk assessment should determine the likelihood of stress resulting in ill-health which will prompt actions to deal with those pressures. For example, a supervisor needs to consider whether an employee has too many competing deadlines before asking that person to do an addition task. The employee needs to clearly understand the pressure they may already be under when deciding whether to refuse the extra work or not.

The risks arising from stress may also vary depending on the task. Some tasks require a higher level of concentration, coordination or strength in order to carry them out safely. As a result, the level of pressure that may be tolerated will vary between industries and occupations.

16.6. STRESS MANAGEMENT SYSTEMS

Stress management systems should only be developed in organisations in which a need is identified. They should be developed in conjunction with grievance procedures to cover both preventative strategies and the provision of help.

Otherwise, the application of good management practices that include appropriate organization of work, a healthy corporate attitude, successfully managed change and an appropriate management style will all contribute towards reducing unwanted stress levels in the workplace. Work organisations need to develop a workplace culture that recognises that job satisfaction factors such as flexibility, autonomy, security, recognition, ownership, participation and involvement are as essential as stable industrial relations. Employers in such organisations should have no doubt that health, safety, security and morale are inextricably linked to employee satisfaction, productivity and customer satisfaction.

16.7. PREVENTION OF STRESS AT WORK

Stress Management Training. Provide stress management training and an employee assistance programme (EAP) to improve the ability of workers to cope with difficult work situations. Stress management programme teach workers about the nature and sources of stress, the effects of stress on health, and personal skills to reduce stress—for example, time management or relaxation exercises. Stress management training may rapidly reduce stress symptoms such as anxiety and sleep disturbances; it also has the advantage of being inexpensive and easy to implement. However, stress management programme have two major disadvantages.

- The beneficial effects on stress symptoms are often short lived.
- They often ignore important root causes of stress because they focus on the worker and not the environment.

Organizational Change: In contrast to stress management training and EAP programmes, reduce job stress by bringing in a consultant to recommend ways to improve working conditions. This approach is the most direct way to reduce stress at work. It involves the identification of stressful aspects of work (*e.g.*, excessive workload, conflicting expectations) and the design of strategies to reduce or eliminate the identified stressors. The advantage of this approach is that it deals directly with the root causes of stress at work. However, managers are sometimes uncomfortable with this approach because it can involve changes in work routines or production schedules, or changes in the organizational structure.

16.8. STEPS TOWARD PREVENTION

Steps 1–Identify the Problem

The best method to explore the scope and source of a suspected stress problem in an organization depends partly on the size of the organization and the available resources. Group discussions among managers, labour representatives, and employees can provide rich sources of information. Such discussions may be all that is needed to track down and remedy stress problems in a small company. In a larger organization, such discussions can be used to help design formal surveys for gathering input about stressful job conditions from large numbers of employees.

Regardless of the method used to collect data, information should be obtained about employee perceptions of their job conditions and perceived levels of stress, health, and satisfaction. The list of job conditions that may lead to stress and the warning signs and effects of stress provide good starting points for deciding what information to collect.

Objective measures such as absenteesim, illness and turnover rates, or performance problems can also be examined to gauge the presence and scope of job stress. However, these measures are only rough indications of job stress-at-best.

Data from discussions, surveys, and other sources should be summarized and analyzed to answer questions about the location of a stress problem and job conditions that may be responsible—for example, are problems present throughout the organization or confined to single departments or specific jobs?

Survey design, data analysis, and other aspects of a stress prevention programme may require the help of experts from a local university or consulting firm. However, overall authority for the prevention programme should remain in the organization.

Step 2–Design and Implement Interventions

Once the sources of stress at work have been identified and the scope of the problem is understood the stage is set for design and implementation of an intervention strategy

In small organizations, the informal discussions that helped identify stress problems may also produce fruitful ideas for prevention. In large organizations, a more formal process may be needed. Frequently, a team is asked to develop recommendations based on analysis of data from Step 1 and consultation with outside experts.

Certains problems, such as a hostile work environment, may be pervasive in the organization and require company-wide interventions. Other problems such as excessive workload may exist only in some departments and thus require more narrow solutions such as redesign of the way a job is performed. Still other problems may be specific to certain employees and resistant to any kind of organizational change, calling instead for stress management or employee assistance interventions. Some interventions might be implemented rapidly (*e.g.*, improved communication stress management training), but others may require additional time to put into place (*e.g.*, redesign of a manufacturing process).

Before any intervention occurs, employees should be informed about actions that will be taken and when they will occur. A kick-off event, such as an all-hands meeting, is often useful for this purpose..

Step 3–Evaluate the Interventions

Evaluation is an essential step in the intervention process. Evaluation is necessary to determine whether the intervention is producing desired effects and whether changes in direction are needed.

Time frames for evaluating interventions should be established. Interventions involving organizational change should receive both short- and long-term scrutiny. Short-term evaluations might be done quarterly to provide an early indication of programme effectiveness or possible need for redirection, Many interventions produce initial effects that do not persist. Long-term evaluations are often conducted annually and are necessary to determine whether interventions produce lasting effects.

Evaluations should focus on the same types of information collected during the problem identification phase of the intervention, including information from employees about working conditions, levels of perceived stress, health problems, and satisfaction. Employee perceptions are usually the most sensitive measure of stressful working conditions and often provide the first indication of intervention effectiveness. Adding objective measures such as absenteeism and health care cost may also be useful. However, the effects of job stress interventions on such measures tend to be less clear-cut and can take a long time to appear.

The job stress prevention programme does not end with evaluation. Rather it should be seen as a continuous process that uses evaluation data to refine or redirect the intervention strategy.

16.9. JOB STRESS AND HEALTH

Stress sets off an alarm in the brain, which responds by preparing the body for defensive action. The nervous system is aroused and hormones are released to sharpen the senses, quicken the pulse, deepen respiration, and tense the muscles. This response (sometimes called the fight or flight response) is important because it helps us defend against threatening situations. The response is preprogrammed biologically. Everyone responds in much the same way, regardless of whether the stressful situation is at work or home.

Short lived or infrequent episodes of stress pose little risk. But when stressful situations go unresolved, the body is kept in a constant state of activation, which increase the rate of wear and tear to biological systems. Ultimately, fatigue or damage results, and the ability of the body to repair and defend itself can become seriously compromised. As a result, the risk of injury or disease escalates.

In the past 20 years, many studies have looked at the relationship between job stress and a variety of ailments, Mood and sleep disturbances, upset stomach and headache, and disturbed relationships with family and friends are examples of stress-related problems that are quick to develop and are commonly seen in these studies. These early signs of job stress are usually easy to recognize. But the effects of job stress on chronic diseases are more difficult to see because chronic diseases take a long time to develop and can be influenced by many factors other than stress. Nonetheless, evidence is rapidly accumulating to suggest that stress plays an important role in several types of chronic health problems-especially cardiovascular disease, musculoskeletal disorders, and psychological disorders.

Early Warning Signs of Job Stress

Headache, sleep disturbances, Difficulty in concentrating, short temper, upset stomach, job dissatisfaction, low morale.

16.10. STRESS, HEALTH, AND PRODUCTIVITY

Some employers assume that stressful working conditions are a necessary evil that companies must turn up the pressure on workers and set aside health concerns to remain productive and profitable in today's economy. But research findings challenge this belief. Studies show that stressful working conditions are actually associated with increased absenteeism, tardiness and intentions by workers to quit their jobs-all of which have a negative effect on the bottom line.

Recent studies of so-called healthy organizations suggest that policies benefiting worker health also benefit the bottom line. A healthy organization is defined as one that has low rates of illness, injury, and disability in its workforce and is also competitive in the marketplace. NIOSH research has identified organizational characteristics associated with both health, low-stress work and high levels of productivity. Examples of these characteristics include the following:

- Recognition of employees for good work performance
- Opportunities for career development
- An organizational culture that values the individual worker
- Management actions that are consistent with organizational values.

IMPORTANT QUESTIONS

1. Define Occupational stress.
2. What are causes of stress.
3. What are various signs of stress.
4. What are duties of employees and employees to minimise stress.
5. What are the ways to measure stress.
6. Explain stress management system.
7. Explain various steps for preventing stress.

❑❑❑

CHAPTER 17

OCCUPATIONAL HEALTH

Health is a state of complete physical, mental and social well being and not merely the absence of disease. According to the joint ILO/WHO committee on organizational health, industrial health is:

(*i*) The promotion and maintenance of physical, mental and social well being of workers in all occupations;

(*ii*) Prevention among workers of ill-health caused by the working conditions;

(*iii*) Protection of workers in their employment from risk resulting from factors adverse to health, and

(*iv*) Placing and maintenance of the worker in an occupational environment adapted to his physical and psychological equipment.

17.1. SIGNIFICANCE OF INDUSTRIAL HEALTH

Workers spend a great deal of their time in industrial settings. In many industries, workers are exposed to various types of health hazards unless the working conditions, lighting, ventilation, cleanliness, temperature space, etc. are proper workers cannot concentrate on work. They do not feel working hard continuously for long time. As a result productivity will be low. Secondly, ill-health will force workers to be absent from work. So the rate of absenteeism and labour turnover is high. The quality of work suffer and the firm's resources will be wasted. In addition, industrial discontent and indiscipline will arise. Thus, industrial health is essential for the following reasons :

(*i*) To minimise absenteeism, and labour turnover.

(*ii*) To reduce industrial indiscipline and accidents.

(*iii*) To improve employee motivation and morale.

(*iv*) To reduce spoilage and cost of operations.

(*v*) To preserve the physical and mental health of employees.

(*vi*) To maintain and improve productivity and quality of work.

Management has a responsibility to ensure that the workplace is free from unnecessary hazards and that conditions surrounding the workplace are not hazardous to employees physical or mental health. Of course, accidents can and do occur on many jobs, and the serverity of these may astound you. There are approximately 20 million work related injuries every year, 390,000 work related illnesses and 100,000 work related deaths. From the turn of the century through the late 1960's remarkable was made in reducing the rate and severity of job related accidents and diseases. Yet the most significant piece of federal registration in the area of employee health and safety was not enacted until 1970.

17.2. OCCUPATIONAL SAFETY AND HEALTH ACT

Because of the increasing complexity and hazardous nature of modern industrial operations, in 1970 the occupational safety and health Act was passed. The OSHA legislation established comprehensive and specific health standards, authorized inspections to ensure the standards are met; empowered the occupational safety and health administration to police the organizations compliance and required employees to keep records of illness and injuries and to calculate accident ratios. The act applies to firms engaged in interstate commerce having one or more employees; approximately 74 million employees are covered. The enforcement agency is the occupational safety and health administration in the Department of Labour. The research agency with respect to industrial hazards is the National Institute for Occupational Safety and Health located in the Department of Health and Human Services. States that submit qualified programmes can take over administration of them receiving a federal grant equal to 50 per cent of their costs; slightly over half have done so.

17.3. WORKING CONDITIONS AFFECTING HEALTH

The factors influencing health are so as follows:

1. Cleanliness: Hygiene is essential to health. Dirt should be removed daily from workplace, furniture, staircases, etc. Properly cleaned and disinfected spittoons must be provided at convenient places. Arrangement should be made for proper disposal of industrial waste. Rats, pests and insects should be destroyed as these are worst carriers of diseases. Drinking water, toilets, restaurant and canteen facilities should be provided for workers.

2. Lighting: Adequate lighting is essential for higher efficiency and good quality of work. Poor quality on the other hand causes eye strain, mental fatigue, accidents and spoilage of materials. A good lighting system should provide (*a*) the right degree of intensity of light depending upon the nature of work (*b*) well-diffused and uniformly spread light all over the work place (*c*) protection from any glare.

3. Temperature and Ventilation: Flow of fresh air with right temperature and humidity is necessary for protection of health, Stale air causes headache to workers. In hot and humid climate, employees feel tired and sleepy. Ventilators, fans, coolers, heaters, air-conditioners help to maintain right temperature and humidity. These also keep the air free from dust, smoke, fumes etc.

4. Freedom from Noise: Too much noise inside and outside the workplace causes disturbance. It does not allow workers to concentrate on the work and their efficiency declines. Internal noise occurs due to conversations, movement of workers and machines, telephone calls etc. Use of sound absorbent materials in walls and ceilings, carpets on the floors, installation of noise producing machines in a separate room, etc., help to minimise such noise. External noise is caused by the movement of persons and vehicles near the workplace. Proper location of factory/ office, double doors, glass penal and sound proof walls may be used to avoid external noise.

5. Dust Control: At some workplaces, there is high ratio of dust. For example, work areas in cotton and jute industries are ladden with dust. Dust affects workers health as well as the life of machinery. Therefore, it is necessary to keep the workplace free from dust. Several measures can be taken for this purpose. Regular cleaning of doors, windows, walls and ceilings is necessary. Floors should be washed daily to avoid accumulation of dust. Electrical installations should be regularly cleaned to maintain the level of lighting, machines, furniture and records should be cleaned regularly.

6. Working Space and Seating Arrangement: Adequate space should be provided for free movement of persons, machines etc. Proper seating is essential for comfort and health of employees. Overcrowding should be avoided as it spoils health and efficiency.

17.4. HEALTH PROGRAMME

The prevention of accidents is a major part of the function of employee maintenance, but constitutes only one segment of a comprehensive programme. The employee's physical condition can be harmed through disease, stress and strain as well as through accidents. It is important for the firm to be concerned with the general health, both physical and mental, of its employees for both economic and humanitarian reasons. Many state workers compensation laws provide for compensation for specific occupational diseases. A major part of the OSHA effort is being directed toward prevention of illnesses issuing from the workplace environment. Moreover poor employee health leads to a high level of absenteeism and a low level of productivity.

Physical Health: There is considerably more industrial activity in the field of safety than there is in the field of health. When health and medical programmes are provided, it has been customary in too many firms to attach this function to safety unit. Too often the health portion of the programme consists solely of a group of medical personnel, who administer physical examinations in the hiring process. The occupational health institute would add to the placement physical examination the following requirement for a properly organized health programme:

(1) A stated health and medical policy.

(2) The performance of periodic physical examinations on all employees exposed to health hazards.

(3) The availability of facilities for voluntary periodic physical examinations for all employees.

(4) A competent medical consulting staff.

(5) Systematic attention to sanitation, safety precautions and industrial hygiene.

(6) A chief medical officer who reports to a responsible member of management.

(7) A well-equipped dispensary for emergency cases and physical examinations.

(8) Properly qualified medical and nursing personnel.

Mental Health: In recent years, an increasing amount of attention has been devoted to industrial mental health. It is difficult to determine how far a business firm should go in such matters as this various state court decisions are assigning liability to employers for emotional stresses engendered by employment.

The rationale for a company mental health programme is the same as for programmes designed to maintain physical health. In the first place, business is responsible for creating many of the tensions and pressures of modern life, which in turn often lead to mental breakdowns. Secondly, mental difficulties of various types are reflected in the firm's productivity and profit records. Mental illness takes its to through alcoholism, high accident rates, high tardiness rates, high labour turnover, and poor human relationships. Studies have shown that the emotionally ill have more accidents and safety violations, file more grievances, and are discharged more often than healthy workers.

Despite the logic of the above, mental health ranks third behind safety and physical health as an area in which private industrial managements expend resources.

The rehabilitation of the alcoholic employee is perhaps the best example of what private firms can do in regard to mental health programmes. It is estimated that 5 to 10 per cent of the national workforce at all levels is affected by a drinking problem. One study indicates that the alcoholic employee was absent 19.4 days annually, as compared with 5.8 days for the nonalcoholic. Because of this cost, plus those issuing form increased accidents and reduced efficiency, more than 50 major corporations conduct rehabilitation programmes in the field.

The recommended procedure for dealing with the alcoholic employee is as follows :

(1) The supervisor maintains accurate records on employee performance, such as productivity, asenteeism, lateness, leaving the workplace, warnings, innovative excuses, and so on.

(2) Though suspecting an alcoholic source, the supervisor talks with the employee only about his or her objective record of performance.

(3) If performance does not change, it is suggested that the employee join the supervisor in meeting with a counselor in the personnel unit.

(4) During the initial phases of this meeting, the supervisor presents the objective performance record and then withdraws.

(5) The counselor "lay sit on the line" and suggests that a discharge is imminent if performance is not improved.

(6) If the employee indicates that alcohol is the source of the performance difficulties, she or he is retained on the payroll and referred to an appropriate outside agency, such as Alcoholics Anonymous, a psychologist, or a hospital specializing in alcoholism treatment.

(7) The counselor follows the case for 1 year after referral to determine the extent of rehabilitation.

Implementation of the above procedure in the Huges Aircraft Company resulted in a success rate of 75 per cent.

A related, but newer, problem is that of employee use of drugs on the job. In a study of 95 subjects in a drug rehabilitation center, 91 reported use of drugs shile on the job, 48 reported they had sold drugs to other employees, and 68 indicated forms of criminal behaviour, including stealing materials, payroll forgery, and so on. Use of heroin was reported by 1.3 per cent, of LSD by 2.6 per cent of methedrine by 2 per cent and of marijuana by 12 per cent.

Though it would appear that the same approach could be taken to drug use as to alcoholism, fewer firms are willing to put forth a significant rehabilitation effort. Some claim that alcoholics do not proselytize for alcoholism they can be helped while keeping them on the job. The drug user often wants to share his or her discovered escape route form the world with other workers. In addition, users may have to sell drugs to finance their personal habit. It is also suspected that the management does not understand the "drug culture" as well as the "alcohol culture". The typical alcoholic is older and has had longer service. The typical drug user is younger and has not built up a service record in the firm. Finally, the illegality of drug usage combined with clinging public attitudes toward certain drugs makes management uncertain about reporting requirements to various government agencies. A not unusual reaction to discovery of a drug user is immediate discharge or an immediate leave of absence without pay.

17.5. FIRST AID

It may be said as the treatment given to a person, before he is taken to hospital. Although in factories, sufficient safety measures are taken to minimise accidents, but accidents cannot be totally avoided and hence proper first-aid facilities for common accidents such as fire, heat stroke, cuts etc., must be provided in every factory by the employers.

A first aid box be provided to a responsible person who must be always available in working hours and he should be trained in first aid. It is necessary that sufficient first aid facilities should exist in all establishments. Some firms have elaborate arrangements of stretchers, ambulance, surgeries etc., in the dispensary. The specified contents of first aid box in workshops employing more than 60 persons as per Factory Act are:

(1) Twenty four small sterilized dressings.
(2) Twelve medium size sterilized dressings.
(3) Twelve large size sterilized dressings.
(4) 10 grams packet sterilized cotton wool.
(5) One snake bit lancet.

(6) One pair of scissors.
(7) Twelve large size burn dressings.
(8) Two (25 gms) bottles of potassium permanganate crystals.
(9) One (100 gms) bottles containing two per cent alcoholic solution.
(10) One (100 gms) bottle of salvotative having the dose and mode of administration indicated on label.
(11) One copy of first aid leaflet.
(12) Twelve rolled bandages 10 cm wide.
(13) Twelve rolled bandages 5 cm wide.
(14) Two rolls of adhesive plaster.
(15) Six triangular bandages.
(16) Two packets of safety pins.
(17) A supply of suitable splits.
(18) One townquet.
(19) Eye drops.

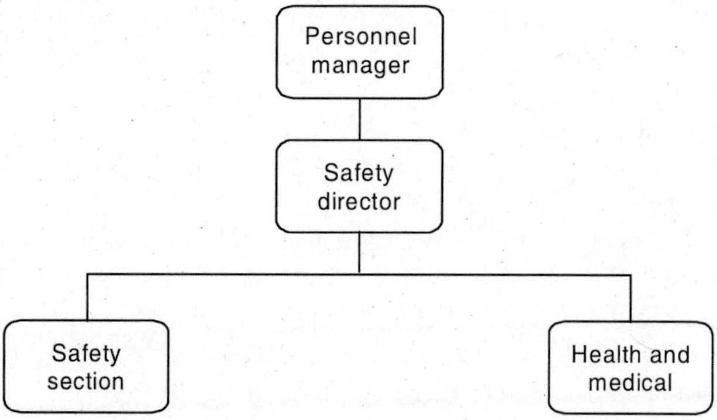

Fig. 17.1. Organization of health and safety.

It is apparent from the list of specifications laid down that only the larger firms would qualify as having properly organized health programmes. A survey of 4.636 plants by the National Institute for Occupational Safety and Health indicated that only 3.1 per cent of all work places provide industrial hygiene services.

In line with both an economic motivation and social responsibility, the larger companies are undertaking a broader "whole person" approach toward industrial health. Some are studying the chronic diseases, such as cancer and heart trouble, and approaching them as a total complex or system involving both industrial and nonindustrial sources. In one company, it is estimated that non-occupational illnesses account for more than 70 per cent of its absenteeism. Payments in the form of sick leave or health fringe benefits may run ten times the amount expended under workers compensation laws. It is maintained that environmental health must stem from *inside* industry, inas much as outside doctors cannot possibly know what goes on inside the plant.

With respect to the field of industrial hygiene, some firms have made an effort to elevate the hygienist to an organization, position superior to that of the safety director. He or she may be placed in a position coordinating safety and medical services. The reasons for this move are several. First, as actually practiced, safety is often backward-looking. The safety engineer learns through mistakes—accidents or diseases. It is quite typical for a serious accident to generate a tremendous amount of activity after the fact. The industrial hygienist is trained to adopt the preventive approach, to spot obvious and incipient dangers, and to request the appropriate speciality to eliminate those dangers. Secondly, a well-qualified industrial hygienist must undego 1 or 2 years of college training in such subjects as physiology, psychology, air analysis, toxicology, and the radiological aspects of industrial health. He or she thus has the background that stimulates analysis and prevention of health and accident hazards.

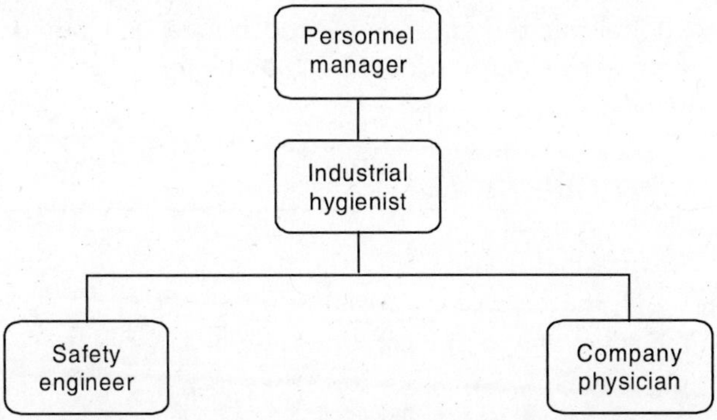

Fig. 17.2. Organization of health and safety.

The industrial hygienist usually analyzes health problems in terms of stress and strains. The sources of occupational stress are four in number: chemical, physical, biological, and social. Occupational stress result in certain strains being placed upon the human body and mind. The total strain is often greater than the sum of the individual stresses. When the stress is too great, the human body or mind breaksdown. Thus, the emphasis is upon the elimination or control of the sources of occupational stress.

Measures for the prevention and control of occupational stress can be applied to the following: (1) the *source* of the stress, (2) the *media* of stress transmission, and (3) the *individual* subjected to the stress. In most health problems, research and correction should be applied in the order given. For example, the industrial hygienist has discovered a machining process that is throwing off a large volume of dust like particles in the air. The first problem is to discover whether the process constitutes a health hazard. One must analyze the size of the particles to determine whether they are small enough to be breathed into the lungs or are large enough to cause them to be coughed out of the throat. One must analyse the rate of fallout of such particles in the various parts of the room. Assuming that a health hazard is found to exist, the first approach is to tackle the source. If the source of the stress can be eliminated, there is no need to restrict the movements and actions of the employee. It may be possible for the process to be changed or for people to be completely eliminated form the offending area.

Falling the elimination of the source of stress, the next approach is to analyze the transmitting media. Can a baffle, screen, or some other type of protection be established to keep the particles away from the persons? Can the air be washed and cleaned? The final step is that of protecting the worker. If neither the source nor the media can be corrected, the human being must be safeguarded. One may be forced to breathe through an oxygen mask.

✎ IMPORTANT QUESTIONS

1. Define Occupational health.
2. Explain the significance of industrial health.
3. Explain the effect of working conditions on health.
4. Explain physical and mental health.
5. Explain the need of first aid and what are contents of first aid box,

❑❑❑

QUESTION BANK

INDUSTRIAL SAFETY AND ENVIRONMENT

(B. Tech., VIIIth Semester)

Q.1. Discuss the relevance of safety in the industry.

Q.2. How plant productivity can be achieved through safety? Explain.

Q.3. What is plant layout? Discuss some of the safety considerations in plant layout.

Q.4. What are different safety measures taken up in an organisation? Discuss.

Q.5. How safety measures affect economy of an industry. Explain.

Q.6. "Employee participation help of achieving safety objectives". Explain.

Q.7. Explain relationship between safety and working environment.

Q.8. Does equipment design affect safety? Explain.

Q.9. Define industrial accident. Discuss the nature, types and causes of industrial accident.

Q.10. What are different measures taken up in an organization to prevent accident? Discuss.

Q.11. What are the methods for assessment of accident cost? Discuss.

Q.12. Define industrial hazard. What are the various hazard identification techniques?

Q.13. Explain accident investigation, reporting and analysis.

Q.14. Define safety planning. Explain purpose, nature, scope and procedure of safety planning.

Q.15. List the range of safety plans.

Q.16. Explain the process of policy formulation and its implementation in safety for the workers.

Q.17. Discuss the various safety standards recommended in an organisation.

Q.18. Discuss the various safety legislation adapted by the organisation.

Q.19. Define environment. What is need of environmental control?

Q.20. List the various environment factors in industry.

Q.21. Explain the effect of temperature, illumination, humidity, noise and vibration on human body and mind.

Q.22. Discuss basics of environment design for improving efficiency and accuracy at work.

Q.23. What are various methods to measure fatigue. How it can be mitigated?

Q.24. Give the formulation for heat load requirement of industrial buildings.

Q.25. What is thermal environment and measurement method.

Q.26. Define thermal comfort. What are thermal limits for comfort? Explain the indices of thermal comfort of heat stress.

Q.27. Explain heat stress and its effect on the efficiency of workers.

Q.28. Explain physiology of heat regulation.

Q.29. What are the various measures taken to control heat exposure in industry.

Q.30. State and explain objectives of ventilation.

Q.31. Differentiate between natural ventilation and mechanical ventilation.

Q.32. Explain Dilution Ventilation.

Q.33. What is air conditioning? How air-conditioners work?

Q.34. What are the various types of air-conditioning systems?

Q.35. Define noise. Discuss continuous and impulse noise.

Q.36. Does noise affect the man is working? Explain.

Q.37. What is noise isolation? What are different noise evaluation procedures? Explain them in detail.

Q.38. What are the various absorption techniques.

Q.39. What are sources of noise? Explain.

Q.40. What are various measures used to control noise?

Q.41. Write a short note on the working of silencers.

Q.42. State the purpose of silencer.

Q.43. Define vibration. Explain the effect of vibration on human body.

Q.44. How vibrations can be measured? Explain.

Q.45. What are the various measures taken to control vibration in industry?

Q.46. What is the purpose of industrial lighting?

Q.47. What are benefits of good illumination.

Q.48. Differentiate between natural light and artificial light.

Q.49. What are various types of artificial lighting? State its objectives.

Q.50. What are principles of good illumination?

Q.51. What are various standards of illumination recommended in an industry?

Q.52. Discuss the maintenance standards needed for lighting and colour so that healthy and conducive environment can be generated for higher productivity.

Q.53. What are considerations to design for lighting industrial areas?

Q.54. How ISO 14000 was introduced?

Q.55. Define EMS. What are the benefits of EMS.

Q.56. What are the key elements of ISO 14001?

Q.57. What are the various environment standards for representative industries.

Q.58. What is ergonomics?

Q.59. Discuss how ergonomics is related to working environment.

Q.60. Discuss the importance of postures while working.

Q.61. What is anthropometry.

❑❑❑

SAMPLE QUESTION PAPER-1

(ME- 402)

INDUSTRIAL SAFETY AND ENVIRONMENT

(B.Tech VIIIth Semester)

Time: 3 Hours] [*Maximum Marks: 60*

Note : *Section A is compulsory. Attempt any* **Four** *questions from Section B and any* **Two** *questions from section C.*

SECTION—A **Marks: 2 Each**

1. (*a*) Discuss the causes of industrial accidents.
   (*b*) What is the nature of planning for safety?
   (*c*) Is safety related to productivity also? Comment.
   (*d*) Define Environment. What is the need of environmental control?
   (*e*) What is meant by Mechanical Ventilation?
   (*f*) List the various types of artificial lighting?
   (*h*) List the various outdoor and indoor sources of noise.
   (*i*) What is the scope of environmental management systems?
   (*j*) Define physical and mental fatigue.

SECTION—B **Marks: 5 Each**

2. Explain the relationship of safety with plant design.
3. How can employees participate in safety measures in a manufacturing organisation?
4. Explain the effect of temperature and humidity on human body and mind.
5. Explain what is meant by Natural ventilation and Dilution ventilation.
6. Why vibrations get developed in work area? Explain their effects and control measures.

SECTION—C **Marks: 10 Each**

7. What is meant by the term Industrial Waste? Explain the various

types of wastes along with the sources that produce them. What is the impact of each on environment?

8. What is meant by Industrial Accidents? Discuss their nature, types and causes in detail.
9. Discuss the various sources and types of artificial lighting. Also discuss the principles of good illumination.

SAMPLE QUESTION PAPER-2

(ME- 402)

INDUSTRIAL SAFETY AND ENVIRONMENT

(B.Tech VIIIth Semester)

SECTION—A

All questions are compulsory **Marks: 2**

1. (*a*) How safety measures affect economy of an industry. Comment.
   (*b*) What is the nature of planning for safety?
   (*c*) Discuss the causes of industrial accidents.
   (*d*) Define environment.
   (*e*) What is thermal comfort and thermal limits for comfort.
   (*f*) What is the purpose of ventilation?
   (*g*) Define continuous and impulse noise.
   (*h*) Explain the effect of vibration on human body.
   (*i*) What are the benefits of good illumination?
   (*j*) What are the different source of industrial waste.

SECTION—B

Attempt any Four **Marks: 5**

2. "Employees participation help on achieving safety objective". Comment.
3. Discuss basics of environment design for improving efficiency and accuracy at work.
4. Explain Local Exhaust Ventilation and Dilution Ventilation.
5. State the purpose of silencer. Explain it working.
6. What are the consideration of design for lighting industrial areas?

SECTION—C

Attempt any Two **Marks: 10**

7. What do you mean by 'industrial accidents'? Discuss their nature, types, causes and prevention in detail.

8. (*i*) What is meant by the term 'industrial waste'? How does it affect our environment? Explain.

   (*ii*) Explain rectification and treatment of industrial waste.

9. What are various environmental standards for representative industries? Discuss the key elements of ISO 14000.

   (*ii*) Define 'Environmental Management Systems' or EMS. What are its benefits.

PREVIOUS YEARS QUESTION PAPERS (SOLVED)

INDUSTRIAL SAFETY AND ENVIRONMENT

(ME-402)

(B. Tech., VIIIth Semester, 2003)

Time: 3 Hours] [*Maximum Marks: 60*

Note : *This paper consists of Three Sections. Section A is compulsory. Answer any* ***Four*** *questions form Section B and any* ***Two*** *questions from Section C.*

SECTION—A **Marks: 2 Each**

1. (*a*) What are the basics of environmental design?

Ans: The environmental design is important as it affects the health, comfort and performance of occupants. The various environmental design factors are:

- noise
- vibration
- temperature
- humidity
- lighting

These factor affect worker's efficiency towards work.

(*b*) How plant productivity can be achieved through industrial safety? Explain.

Ans: Safety and productivity are closely linked with each other. If a worker is mentally ensured that he is working in a safe working condition and his life and limbs are safe in all possible manner and his working zone is free from all types of hazards then in such situations the morale of the worker will be high and he will possibly work with more zeal and thus the productivity will increase considerably.

(*c*) Give the formulation for heat load requirement of Industrial Buildings.

Ans: The commonly used measure of overall heat load in the workplace is the wet bulb globe temperature (WBGT) index. The wet bulb temperature is calculated using a formula that takes into account air temperature, speed of air movement, radiant heat from hot objects, sunshine and body cooling due to sweat evap.

- Four outdoors with direct sun exposure:

$$WBGT = 0.7 \times Temp._{wet\ bulb} + 0.2 \times Temp._{globe} + 0.1 \times Temp_{.air}$$

- For indoors or outdoors without direct sun exposure:

$$WBGT = 0.7 \times Temp._{wet\ bulb} + 0.3 \times Temp._{globe}$$

where: $Temp._{wet\ bulb}$ natural wet bulb temperature measured by using a thermometer whose bulb is covered with wet cotton cloth and is cooked by the natural air movement.

$Temp._{globe}$ temperature measured using a black globe thermometer.

$Temp._{air}$ temperature measured using a conventional thermometer.

All the temperature are to be expressed in '*C*'.

(*d*) How ergonomics can be applied in improving the industrial safety.

Ans: Ergonomics is a multi-disciplinary science which provides the scientific basis for optimized design of the work environment compatible with capabilities and limitations of the working populations. An ergonomist is trained to recognize, evaluate that result from a poor fit between the worker and the workplace, these hazards may result in acute injury, chronic musculoske detail disorders, or mental/psychosocial stress.

(*e*) List down the reasons of Industrial Accidents.

Ans: A large number of seasons lead to industrial accidents. Basic factors responsible for causing accidents are unsafe acts of the employees, unsafe working conditions and last but not the least the chance occurrence.

(*f*) What are the different sources of industrial waste?

Ans: The various sources of industrial waste are:

Liquid waste: It is the waste water resulting from industrial processing operations. All the "wet processing" industries such as those engaged in manufacturing of food products, fruits, sugar industries, dairy, paper and pulp manufacturing, number of chemical industries, laundries consume large quantity of water and produce wate in large quality.

Solid Waste: The solid waste may result wood working industries (saw dust) the fragment of metal fabrics plastics etc.

Gaseous waste: In some industries nuisance or poisnous gases, fumes and dusts are produced. For example: melting furnaces, foundry, chemical industries, welding, smithy, refineries, power stations (thermal) etc.

(*g*) Does noise affect the man's working? Explain.

Ans: Noise is a serious environmental problem and a health hazard. It is not merely a nuisance. It has far reaching consequences and have many physical, physiological as well as psychological effects on human being.

(*h*) Explain dilution ventilation.

Ans: Dilution ventilation is usually applied to the control of contaminants in situation meeting the criteria:

- To control vapours from low toxicity solvents.
- To control contaminants released over such a large area on in such a manner that local exhaust ventilation is impossible, impractical or expensive.
- Small quantities of contaminates released into the work room.
- No corrosion or other problems from the diluted contaminats in the work room air.
- Sufficient distance from the worker to the contaminants source to allow dilution to safe levels.

(*i*) What is thermal environment and measurement method?

Ans : Thermal environment is related to the combined effects of radiant temp., humidity of air, velocity etc. An undesirable thermal environment results in significant discomfort. The most commonly used measure in the workplace is the wet bulb globe temperature (WBGT) index. The wet bulb globe temperature (WBGT) is calculated by using the following equations :

- For outdoors with direct sun exposure:

$$WBGT = 0.7 \times \text{Temp.}_{\text{wet bulb}} + 0.2 \times \text{Temp.}_{\text{globe}} + 0.1 \times \text{Temp.}_{\text{air}}$$

- For indoors or outdoors without direct sun exposure:

$$WBGT = 0.7 \times \text{Temp.}_{\text{wet bulb}} + 0.3 \times \text{Temp.}_{\text{globe}}.$$

Where: $\text{Temp.}_{\text{wet bulb}}$ natural wet bulb.

$\text{Temp.}_{\text{globe}}$ temperature measured using a black globe thermometer.

Temp_{air} temperature measured using a conventional temp. thermometer. All temperatures are to be expressed in C.

(*j*) Differentiate between noise and vibration.

Ans: A regular periodic variation in value about mean is called vibration. However noise results when irregular vibrations are made at irregular interval.

SECTION—B (5 marks each)

2. What are the methods used for assessment of accident cost? Discuss.

Ans: Accidents are more expensive than most people realize because of hidden cost. Some cost are obvious for example, workers's compensation claims which cover medical costs and indemnity payments for an injured or ill workers. These are the direct cost of accidents.

But the cost to train and compensate a replacement worker, repair and damaged property investigate the accident and implement corrective action and to maintain insurance coverage are indirect cost. Even less apparent are the cost related to schedule delays, added administrative the time, lower morale increased absenteeism and poor customer relations.

These are also indirect cost.

The estimation of annual cost of accidents at workplace involves the following steps:

Step I: Estimated Annual Accident costs

Annual number of occupational deaths = ______ × *$ 910,000 = $ ______

Annual number of lost workday cases = ______ × Rs * $1000 = $ ______

Annual number of repeatable cases without lost work days = ______ × $*7,000 = $ ______

Total estimated annual cost of occupation deaths, injuries and illness = $ ______

* Using National safety council average costs for 1998, includes both direct and indirect costs, excludes property damage.

Step II: Impact of accidents on profit and sales. (Compare accident costs to company profits)

Sales volume = $ ______

Profit margin = ______ %

Sales volume × % profit = Annual profits = $ ______

Accident cost (Estimated or actual) = & ______

Accident costs as a percent of profit = ______ %

Amount of sales Needed to Replace lost Profits = $ ______

(If profit margin is 5 %, then it takes $ 20 of sales to replace every dollar of loss)

3. Discuss different basic requirement needed for improving efficiency and accuracy at work.

Ans: Environmental factors have influence the physical well being and therefore efficiency of worker towards his work. Some of the environment factors like illumination, ventilation, temperature, noise and vibrations can be very often important factors in determining efficiency of workers because these factors may either lead to fatigue or strain or may create conditions that may attenuated the effects of fatigue, through these may not cause it.

Noise is an important consideration at every stage. A noisy site, for example, will strongly affect the over-all design of a building and may necessitate costly noise control measures.

Other requirements for increasing the efficiency of worker are solar load and solar penetration during working hours, day time temperatures, humidity high rainfalls in some areas.

Another important environment factor which affect worker's efficiency is lighting or illumination. Improper illumination can cause accidents, discomfort to workers, defects in the eyesight of the workers in the long run.

Temperature at the workplace also have influence on workers performance. Both very high and very low temperatures make workers feel uneasy, fall ill and thereby adversely affecting their performance.

Air quality at workplace also affects the workers performance. The replacement of foul air by fresh air is achieved by ventilation. It is achieved by providing enough openings at convenient locations so that natural wind can pass through the space and automatically replace the foul air. The distribution of windows, doors and openings, their number, the cell height as well as orientation of the structure need careful consideration.

4. State and explain objectives of ventilation.

Ans: The objective of ventilation is to maintain the right exchange of gases in the body. This occurs by supplying tissues with oxygen (O_2), a gas needed for metabolic processes and eliminating carbon dioxide (CO_2) which accumulates as a result of processes generating energy.

Ventilation becomes necessary where people stay indoors because human body consumes oxygen and releases carbondioxide. In a closed room without ventilation, the oxygen content in the breathing air will diminish and cause discomfort. The contimination of air by sweating and body odours also necessities ventilation.

In factories many working together itself causes air pollution. Apart from the human body, machines and processes contribute to heat. Dust, gases and fuels generated in the process also make the work environment more uncomfortable and sometimes dangerous.

So the purpose of ventilation is to improve the air quality around the worker. Also to provide thermal comfort by maintaining adequate heat balance of the body and to get rid of contamination produced in the various processed of the industry concerned.

5. Discuss the maintenance standards needed for lighting and colour so that healthy and conducive environment can generated for higher productivity ?

Ans: The maintenance standards relating to lighting and colour are:

- Fluorescent fittings equipped with electric ballasts and connected to dimming control equipment must be treated as systems rather than individual fittings. Tube replacement should be undertake on a "whole room" basis at periodic intervals, as demanded by effective tube life.
- Fittings must be planned to allow simple replacements.
- Acrylic diffusers, with built-in silicon gaskets prevent loss of gaskets during lamp change.
- Stainless steel diffuser dips to be used for ease of maintenance.
- Gear tray and diffuser to be attached to body via a lanyard for ease of installation.
- Regular cleaning and prompt replacement of lamp is essential in any well operated industrial lighting system. Analyse luminous construction and reflector finish.
- In general, light shades for colouring surfaces is recommended in factories. The colour schemes should be such that a proper harmony is maintained. However just for contrast machine with moving parts should be painted with dark colours to distinguish it from the stationary pars of the machines.
- Safety colours are helpful for identifying physical hazards as well as reproducing fire fighting operations. Each colour indicates a specific purpose, *e.g.,*

  Red: Fire prevention and fighting.

  Orange: Those points of machinery which could cause injury.

  Yellow: General caution as for example against falling etc.

6. What is noise isolation? Explain them in detail noise evaluation procedures? Explain them in detail.

Ans: The most direct approach for noise control is to reduce the sound produced by the source itself. It is called noise isolation. Vehicles and machinery can be effectively muffled to reduce the noise . In industrial facilities, noise reduction can be obtained by using rigid, sealed enclosure around machinery, acoustric absorbing material can live the inside of the enclose. Machines or their enclosures can be isolated from the floor using special spring mounts or absorbent mounts and pads.

One of the best but often overlooked, methods of noise isolating is regular and through maintenance of operating machinery.

Noise Evaluation Procedure:

1. Identification of Hazardous Noise Areas: The work areas where noise levels equal or exceeds 85 dBA are identified. Personnel who work in these areas shall have hearing protection supplied to them, shall be instructed in its proper use and be required to wear this equipment when in these identified areas. It is the responsibility of the areas supervisior to ensure that these precautions are maintained.

2. Noise Measurement and Exposure Assignment: The monitoring of employees for noise exposure is made up of two parts, area and personal monitoring. Area measurement are generally obtained first. If noise levels are at or above 80 dBA, personal monitoring using dosimeters is then performed. Sample data sheets will be used to record monitoring data for both area and personal noise monitoring results.

3. Area Measurement: In an area survey, measurement of environmental noise levels are recorded using a sound level meter to identify work areas where employee's exposures may be above hazardous levels and where more through exposure monitoring may be needed.

4. Personnel Monitoring: Each employee to be monitored will have a dosimeter placed on him/her at the beginning of his/her normal work shift with the microphone placed in the *"hearing zone"*. The dosimeter will be worn for the whole shift. At the end of the work shift it is to be removed and information printed area.

5. Remonitoring of Hazardous Noise Areas: All area where noise levels equal or exceed 80dBA shall be remonitored at least every two years.

6. Remonitoring Due to charges: Any area with noise levels that equal or exceed 80 dBA shall also be remonitored whenever a charge production process, equipment or control increase the noise exposure to noise levels at or above 80 dBA on a time weighted average basis

SECTION—C

7. (*i*) Discuss the different type of industrial waste? How does it affect our environment? Explain.

(*ii*) What are different industrial wastes of a process industry. Give its reasons and rectification.

Ans . (*i*) Refer Article 11.3 on page .... (Chapter 11)

(*ii*) Refer Article 11.2 , 11.5, 11.6 ..... (Chapter 11)

8. (*i*) What are different safety measures taken up in a manufacturing organisation? Discuss.

(*ii*) How the worker participation can help any organisation in building its environment as well as its goal achieving ? Discuss.

Ans: (*i*) Refer Article 1.7 on page ..... (Chapter - 11)

(*ii*) Refer Article 1.3 on page .... (Chapter - 1)

9. (*i*) What are various types of artificial lighting. State its objectives.

(*ii*) Explain the process of policy formulation and its implementation in safety for the works.

Ans: (*i*) Refer Article 10.6 on page ..... (Chapter 10)

(*ii*) Refer Article on page .... (Chapter - 2).

❑❑❑

INDUSTRIAL SAFETY AND ENVIRONMENT

(ME-402)

(B. Tech., VIIIth Semester, 5001)

Time: 3 Hours*]** **[*Maximum Marks: 60

Note : *This paper consists of Three Sections. Section A is compulsory. Answer any* ***Four*** *questions form Section B and any* ***Two*** *questions from Section C.*

SECTION—A **20 Marks**

1. (*a*) Explain the relationship of safety with plant design.
   (*b*) Describe the nature of planning for safety.
   (*c*) How safety is related to economics?
   (*d*) What is the effect of illumination on human body and mind?
   (*e*) Describe, what do you understand by local relief.
   (*f*) What are the principles of good illumination?
   (*g*) How noise can be measured? Explain.
   (*h*) Discuss the concept of ISO 14000.
   (*i*) Describe various types of wastes.
   (*j*) What do you understand by environmental standards for representative industries?

SECTION—B **5 Marks each**

2. Describe the impact of industrial wastes on environment. How this impact can be reduced?
3. Define planning for safety. Discuss its purpose, scope and procedure.
4. Discuss the role of employee's participation in safety measures. How manufacturing organisation is responsible for safety measures?
5. Discuss the effect of noise on man. How noise and vibrations are related? How vibration is measured?
6. Describe the nature, causes and types of industrial accidents. How these can be prevented.

SECTION—C **10 Marks each**

7. (*a*) Explain the meaning of environment and discuss the necessity

of environmental control. Also explain how mental fatigue can be measured.

(*b*) Discuss the effect of temperature and humidity on human mind. 5, 5

8. Describe the following:

(*a*) Control of radiant heat.

(*b*) Thermal limits for comfort.

(*c*) Control of heat exposure.

(*d*) Physiology of heat regulation. 2½×4

9. (*a*) What are the benefits of good illumination? What are the recommended optimum standards of illumination? Explain.

(*b*) Describe the maintenance standards relating to lighting and colour. 5, 5

❏❏❏

INDUSTRIAL SAFETY AND ENVIRONMENT

(ME-402)

(B. Tech., VIIIth Semester, 2123)

Time: 3 Hours] [*Maximum Marks: 60*

Note : *This paper consists of Three Sections. Section A is compulsory. Answer any* ***Four*** *questions form Section B and any* ***Two*** *questions from Section C.*

SECTION—A Marks: 2 Each

1. (*a*) What is the Meaning of Safety?
   (*b*) How safety and ergonomics are related? Explain.
   (*c*) What are the Indices of heat stress?
   (*d*) Differentiate between continuous and impulse noise.
   (*e*) What is local relief?
   (*f*) Define solar architecture.
   (*g*) What is the effect of humidity on human body?
   (*h*) State the objectives of good illumination.
   (*i*) Differentiate between Natural and Mechanical ventilation.
   (*j*) Explain worker participation in safety.

SECTION—B Marks: 5 Each

2. Discuss different noise absorption techniques.
3. What are the recommended optimum standards of illumination? Explain.
4. What are Industrial Hazards? Givè the Hazard rectification techniques.
5. What are the methods that are used to measure the physical and mental fatigue? Discuss.
6. Explain why Industrial accidents occur. Give their reasons and how they can be eliminated. Suggest.

SECTION—C Marks: 10 Each

7. Explain the following terms:
   (*a*) Accident investigation.

(*b*) Effect of temperature on human body and mind.

(*c*) Benefits of good illumination.

(*d*) Measurements of vibrations.

(*e*) ISO 14000.

8. (*a*) Explain the concept of Air conditioning process ventilation. Mention the human comfort condition clearly and discuss.

   (*b*) Does lighting effect the working of human being. It so, explain it in detail in the context of higher productivity.

9. Explain Industrual waste and its methods of treatment and disposal. Give the examples of Industries which are source of higher industrial waste. Suggest the methods for minimization of waste.

❑❑❑

INDUSTRIAL SAFETY AND ENVIRONMENT

(ME-402)

(B. Tech., VIIIth Semester, 2121)

Time: 3 Hours] [*Maximum Marks: 60*

Note : *This paper consists of Three Sections. Section A is compulsory. Answer any* ***Four*** *questions form Section B and any* ***Two*** *questions from Section C.*

SECTION—A **Marks: 20**

1. (*a*) What can be causes of Industrial Accidents?
   (*b*) What do you know about HAZOP and HAZAN?
   (*c*) What is the need of planning for safety?
   (*d*) How is safety related to Economics?
   (*e*) What is the harmful effect of noise on human body?
   (*f*) How can the heat exposure be controlled?
   (*g*) Give *three* benefits of good illumination.
   (*h*) How do silencers control the noise?
   (*i*) What do you know about whole body vibrations?
   (*j*) How does industrial wastes affect the environment?

SECTION—B **Marks: 5 Each**

2. How can the safety be integrated with equipment and plant design? What kind of accidents be avoided in this manner?
3. What kind of safety plans can be organisation have? How does it implement the same?
4. What factors do govern the quality of visual environment? Discuss briefly.
5. How does human body respond to vibrations?
6. What can be different types of industrial wastes? Write different methods of its treatment and disposal.

SECTION—C **Marks: 10 Each**

7. (*a*) What techniques of hazard identification are known to you? Explain in detail.

(*b*) How are the accidents investigated? How is the cast associated to accidents assessed? 5, 5

8. (*a*) What is the purpose of ventilation in industry? How does it affect the human performance?

   (*b*) How does the illumination level affect the performance of manual activities? Give the recommended levels of illumination for some general type of industrial tasks and situations.

9. (*a*) Write down the various clauses of ISO-14000 standard.

   (*b*) How does the environmental standard ensure safe and healthy environment around?

INDUSTRIAL SAFETY AND ENVIRONMENT

(ME-402)

(B. Tech., VIIIth Semester, 2056)

Time: 3 Hours*]** **[*Maximum Marks: 60

Section A contains 10 Sub-Questions. Write all the questions each carries 2 marks.

Section B contains 5 questions. Write any four. Each carries 5 marks.

Section contains 3 questions. Write any two. Each carries 10 marks.

SECTION—A **Marks: 20**

1. (*a*) List various causes of Industrial accidents.
   (*b*) Mention the planning methods employed for safety.
   (*c*) Describe the relationship between Industrial Revolution and safety legislation.
   (*d*) Explain briefly the effect of fatigue on accidents.
   (*e*) Why is ventilation important at shop floors?
   (*f*) Specify the Industrial lighting and mention the benefits of good illumination.
   (*g*) How is sound levels are mentioned and list the levels of sound in various industries?
   (*h*) List various industrial wastes and their methods of treatment.
   (*i*) Explain the term Ergonomics.
   (*j*) Briefly write about safety features incorporated in ISO 14000.

SECTION—B **(10×2=20 Marks)**

2. Safety in industry today is largely a matter of complying with government regulation. Justify.
3. Describe the various safety measures and their contribution in augmenting plant productivity.
4. What is a common breakdown in implementation of injury prevention procedures? Suggest corrective measures.
5. Design the ventilation at work by considering all factors.

6. Briefly explain vibrations at a shop floor and enunicate the methods of reducing.

SECTION—C **Marks: 10 Each**

7. Describe a major industrial accident occurred in India and mention the rules and guidelines flouted with regard to safety.
8. Explain some of the factors that complicate the determination of hearing loss.
9. Briefly explain the occupational health hazards and the safety procedures to be adopted in industries.

❑❑❑

INDUSTRIAL SAFETY AND ENVIRONMENT

(ME-402)

(B. Tech., VIIIth Semester, 2054)

Time: 3 Hours] [*Maximum Marks: 60*

Note: *Section A is compulsory. Answer any **Four** questions form Section B and any **Two** questions from Section C.*

SECTION—A **Marks: 20**

1. (*a*) Discuss the causes of industrial accidents.
   (*b*) What is the nature of planning for safety?
   (*c*) Is safety related to productivity also? Comment.
   (*d*) Define Environment. What is the need of environmental control?
   (*e*) What is meant by Mechanical Ventilation?
   (*f*) List the various types of artificial lighting.
   (*g*) List the various outdoor and indoor sources of noise.
   (*h*) Why disposal of Industrial waste is very critical?
   (*i*) What is the scope of environmental management systems?
   (*j*) Define physical and mental fatigue.

SECTION—B **Marks: 5 Each**

2. Explain the relationship of safety with plant design.
3. How can employees participate in safety measures in a manufacturing organisation?
4. Explain the effect of temperature and humidity on human body and mind.
5. Explain what is meant by Natural ventilation and Dilution ventilation.
6. Why vibrations get developed in work area? Explain their effects and control measures.

SECTION—C **Marks: 10 Each**

7. What is meant by the term Industrial Waste? Explain the various

types of wastes along with the sources that produce them. What is the impact of each on environment?

8. What is meant by Industrial Accidents? Discuss their nature, types and causes in detail.

9. Discuss the various sources and types of artificial lighting. Also discuss the principles of good illumination.

❑❑❑

INDUSTRIAL SAFETY AND ENVIRONMENT

(ME-402)

(B. Tech., VIIIth Semester, 2125)

Time: 3 Hours] [*Maximum Marks: 60*

Note: *Section A is compulsory. Answer any* ***Four*** *questions form Section B and any* ***Two*** *questions from Section C.*

SECTION—A **2×10=20**

1. What is ISO 14000?
2. What is audiometer, sound level meter and noise dose meter?
3. Enumerate the list of Industrial waste.
4. What is the SI unit of luminous intensity?
5. Define sound power (Lw)
6. When does grease become a hazard to safety?
7. How should chips be removed from the table or bed of a machine?
8. Write the name of light measuring instruments.
9. What do you understand by guarding of machines?
10. Write about the fire prevention methods.

SECTION—B **5 × 4=20**

1. Discuss the impact of industrial waste on Environment.
2. Write about EMS documentation.
3. Discuss the different methods used to control the noise in Industries.
4. What factors you will keep in your mind during the designing of lighting system for industry?
5. What are the liminations and importance of ventilation system for industry?

SECTION—C **10×2=20**

1. Discuss about the factory act amendment in 1987 and their implications.

2. (*a*) Discuss the general shop safety points in detail.

   (*b*) What will be the sound pressure level in decibel if rms SPL is 2 N/m^2?

3. Discuss the different environment factors (temperature, illumination, humidity, noise and vibratino) and their Effect on human body and mind.

❑❑❑

INDEX

□□□